Mathematics Elsewhere

Mathematics Elsewhere

An Exploration of Ideas Across Cultures

MARCIA ASCHER

Princeton University Press
Princeton and Oxford

Published by Princeton University Press,
41 William Street, Princeton, New Jersey 08540

In the United Kingdom: Princeton University Press,
3 Market Place, Woodstock, Oxfordshire OX20 1SY

Third printing, and first paperback printing, 2005
Paperback ISBN 0-691-12022-6

The Library of Congress has cataloged the cloth edition of this book as follows
Ascher, Marcia.
Mathematics elsewhere : an exploration of ideas across cultures / Marcia Ascher.
p. cm.
Includes bibliographical references and index.
ISBN 0-691-07020-2 (alk. paper)

2002108312

British Library Cataloging-in-Publication Data is available

This book has been composed in Times by Deerpark Publishing Services

Printed on acid-free paper. ∞

pup.princeton.edu

Printed in the United States of America

5 7 9 10 8 6 4

ISBN-13: 978-0-691-12022-5 (pbk.)

To Bob

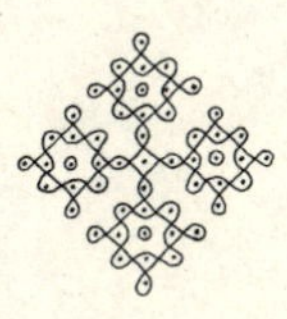

Contents

amazon.com

8253

SDy40cHwZN

Returns Are Easy!

Visit http://www.amazon.com/returns to return any item - including gifts - in unopened or original condition within 30 days for a full refund (other restrictions apply). Please have your order ID ready.

Your order of December 14, 2011 (Order ID 002-1875893-3541051)

Qty.	Item	Item Price	Total
1	**Designing an Anthropology Career: Professional Development Exercises** Briller, Sherylyn H. --- Paperback **(** P-3-C49F114 **) 0759109435**	$18.80	$18.80
1	**Mathematics Elsewhere: An Exploration of Ideas Across Cultures** Ascher, Marcia --- Paperback **(** P-1-B28E26 **) 0691120226**	$28.49	$28.49

We've sent this part of your order to ensure quicker service. The other items will ship separately.

Subtotal	$47.29
Shipment Total	$47.29
Paid via credit/debit	$47.29
Balance due	$0.00

Have feedback on how we packaged your order? Tell us at www.amazon.com/packaging.

926/Dy40cHwZN/-2 of 2-//CPS/second/6842203/1214-23:00/1214-11:31/sp074888253/1-1

1A3

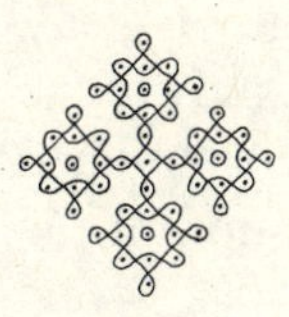

Preface

The book is about some mathematical ideas of people in traditional or small-scale cultures; that is, it elaborates several specific cases of mathematical ideas and their cultural embedding. The use of extended examples is crucial as it gives substance to the ideas involved, including their actual expressions and integration in particular cultural settings. The book is intended as another step toward a global and humanistic history of mathematics.

No special knowledge is assumed of the reader. The book is for mathematicians, as well as non-mathematicians, who are curious about mathematical ideas and about peoples in other cultures; for educators as well as their students; and for anyone interested in the history of mathematics.

My earlier book, written some ten years ago, was similar in intent and in the emphasis on extended case studies. In a sense, this book and the earlier one complement each other; each discusses different mathematical ideas, different cultures, and different cultural contexts. Here, for example, I discuss the structuring of time and the logic of divination, while there I included the organization of space and number words and number symbols.

The interest shown in my work encouraged me to continue my explorations into the mathematical ideas of traditional peoples. I particularly wish to thank those people who invited me to discuss my research at their colleges, at sessions of national mathematics meetings, and at such special events as the NSF/MAA sponsored Institute in the History of Mathematics and Its Use in Teaching (1998) and the 33rd History of Mathematics Conference at Oberwolfach, Germany. Each of these occasions helped me to refine my thoughts and develop means of conveying them to others.

I am indebted to my colleague Karine Chemla and to Serge Pahout for bringing my earlier work to the attention of French readers, and to Claudia Zaslavsky for her ongoing friendship and exchange of ideas. Ithaca College provided the reduction of one course for my study of the Marshall Island stick charts (discussed in Chapter 4) and continued after I became an Emerita Professor to support the preparation of the manuscript of this book.

Although they are acknowledged in the text, I thank those who permitted me to reproduce figures, photographs, or quotations. Special thanks to Congressman Maurice D. Hinchey (26th Congressional District) for helping me obtain permission to reproduce a U.S. postage stamp, and to Dorothy Owens for her help in preparing this book. My appreciation also goes to all of those who helped by reading parts of the manuscript, suggesting further sources, and questioning and commenting on what I wrote or what I said.

Above all, I thank my husband, Robert Ascher, for his encouragement, critical reading, and constructive suggestions.

Marcia Ascher

Mathematics Elsewhere

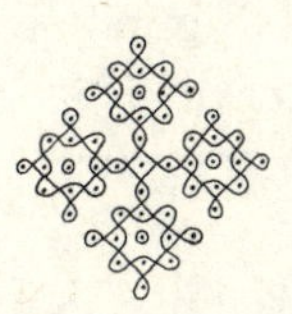

Introduction

As we move into the twenty-first century, we are ever more aware that we are connected to people in other cultures throughout the world. Through expanding communication networks and spreading markets, more experiences of ours and theirs are becoming similar. But, at the same time as we move toward greater likeness, we realize that there is much that we do not and did not share. In particular, we have come to understand that different cultures have different traditions and different histories. Even the same or similar happenings had different effects and different meanings when integrated into different cultural settings and interpreted through different cultural lenses.

This is just as true for mathematical ideas as it is for other aspects of human endeavors. Different cultures emphasized different ideas or expressed similar ideas in different ways. What is more, because cultures assort or categorize things differently, the context of the ideas within the cultures frequently differ.

Among those who study and write about the history of mathematics, there has been growing understanding that what is generally referred to as modern mathematics (that is, the mathematics transmitted through Western-style education) is, itself, built upon contributions from people in many cultures. There is now greater acknowledgment of, in particular, mathematical developments in China, India, and the Arabic world. In addition, there is increased recognition of the work of individuals from an expanding diversity of backgrounds.

There are, however, still other instances of ideas that did not feed into or effect this main mathematical stream. This is especially true of occurrences in traditional or small-scale cultures. In most cases, these cultures and their ideas were unknown beyond their own boundaries, or misunderstood when first encountered by outsiders. During the past 80 years, there have been vast changes in theories, knowledge, and understanding about culture, about language, and about cognitive processes. Yet, only recently have these newer understandings started to impinge upon histories of mathematics to modify the earlier, long-held and widespread, but, nonetheless, erroneous depictions of traditional peoples.

First and foremost, we now know that there is no single, universal path—following set stages—that cultures or mathematical ideas follow. With the exception of specifically demonstrated transmissions of ideas from one culture to another, it is assumed that each culture developed in its own way. When we introduce the varied and often quite substantial mathematical ideas of traditional peoples, we are *not* discussing some early phase in humankind's past. We are, instead, adding pieces to a global mosaic. In terms of our picture of *global* history, we are supplying complexity and texture by incorporating expressions from different peoples, at different times, and in different places. We are, in short, enlarging our understanding of the variety of human expressions and human usages associated with the same basic ideas.

Our focus, then, is elaborating the mathematical ideas of people in these lesser known cultures, that is, the ideas of peoples in traditional or small-scale cultures. In an earlier work, some of the peoples whose mathematical ideas I introduced were the Inuit, Iroquois, and Navajo of North America; the Incas of South America; the Caroline Islanders, Malekula, Maori, and Warlpiri of Oceania; and the Bushoong, Kpelle, and Tshokwe of Africa. Here we continue to enlarge our global vision by discussing, among others, ideas of the Maya of South America; the Marshall Islanders, Tongans, and Trobriand Islanders of Oceania; the Borano and Malagasy of Africa; the Basque of Europe; the Tamil of southern India; and the Balinese and Kodi of Indonesia. Each of these instances adds to our knowledge, but at the same time, makes us all the more aware that it is only a beginning: It is estimated that about 5000–6000 different cultures have existed during just the past 500 years. We will never know about the ideas of those that no longer exist, but there are several hundred that we can know more about.

There is no single, simple way to define a culture. In an attempt to capture all of its nuances, there are many different definitions. By and large, however, the definitions have in common that a culture is a group that continues through time, sharing and being held together by language, traditions, and mores, as well as ways of conceptualizing, organizing, and giving meaning to their physical and social worlds. Often it is associated with a particular place. To say that a culture continues through time is not to say that it is static. All cultures are ever-changing. What varies, however, is the pace of change. In general, traditional or small-scale cultures, as contrasted with, say, post-industrial societies, are more homogeneous and slower to change. Today, throughout the world, there is an overlay of a few dominant cultures, and no culture has remained unmodified by its contacts with others. Nevertheless, traditional cultures still exist, even if sometimes alongside of, or even within a dominant culture.

Where traditions changed slowly or persisted for a long time, we speak about them using the conventional idiom of "the ethnographic present," that is, we describe them at some unspecified time when the traditional culture held sway. However, we will, where we can, note the time depth of the tradition described, and cite some of the ways it has been modified or adapted, while, nevertheless, persisting to varying degrees in its underlying coherence. We will even discuss how a tradition that has been ongoing for hundreds of years both continues in its familiar form and yet becomes involved with a newly developed technology that has been introduced.

Although most of us have a notion of what *mathematics* is, the term has no clear and agreed upon definition. Expansion of the term generally relies on citing examples from one's own experience. To incorporate the ideas of others, it is necessary to clarify our definition and to move beyond the contents of the familiar settings of mathematics, that is, to look beyond the classroom and beyond the work of professional mathematicians. We will, therefore, speak instead of the more inclusive *mathematical ideas*. And, we will, first of all, specify what we take these to encompass: Among mathematical ideas, we include those ideas involving *number, logic, spatial configuration, and, more significant, the combination or organization of these into systems and structures.*

Most cultures do not set mathematics apart as a distinct, explicit category. But with or without that category, mathematical ideas, nonetheless, do exist. The ideas, however, are more often to be found elsewhere in the culture, namely, integrated into the contexts in which they arise, as part of the complex of ideas that surround them. The contexts for the ideas might be, for example, what we categorize as navigation, calendrics, divination, religion, social relations, or decoration. These, in fact, are some of the contexts for mathematical ideas that we will elaborate here. As we discuss the ideas, we also discuss their cultural embedding. Were we to present the ideas divorced from their contexts, they might look more like our own modern mathematics. This approach, however, would distort a major difference—most practitioners of modern mathematics value their ideas because they believe them to be context-free; others value their ideas as inseparable from the cultural milieu that gives them meaning.

Just as most cultures do not have a category called mathematics, they do not group mathematical ideas together as we do—that is, their ideas are not neatly partitionable into, say, algebra, geometry, model building, or logic. The extended examples that we discuss will determine which ideas are presented and the way they are grouped together.

In the chapters that follow, although we discuss the mathematical ideas of others, we do, nevertheless, view them from within our own cultural and mathematical frameworks. For understanding, we call upon similar ideas and concepts we have learned, and we use the vocabulary we share with the reader to convey our understanding. As outsiders to these cultures, we cannot do otherwise. It may well be that other cultures have some ideas too dissimilar from our own for us to detect, just as we have some ideas they do not have. What is crucial, however, is that we not impute to others ideas and concerns that are our own, and that we not be constrained by prejudgments. The process of viewing the ideas of others may lead us to think in more detail about some of our own ideas. In particular, it may lead us to identify some of our unstated assumptions. We may, perhaps, find that some ideas we have taken to be universal are not, while other ideas we believed to be exclusively our own, are, in fact, shared by others.

CHAPTER 1

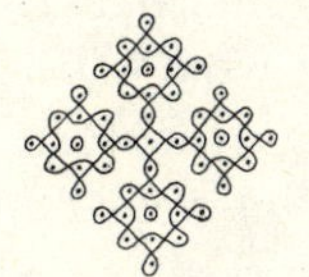

The Logic of Divination

1 Divination, in one form or another, has at some time been practiced in almost every culture. In general, it is a decision-making process, utilizing, as part of the process, a randomizing mechanism. The decisions coming out of the process sometimes involve the determination of the cause of an event or, more often, how, when, or whether to carry out some future action. In different cultures, and at different times, the randomizing mechanisms have varied considerably, involving animals or animal parts, lots or dice, sticks, or whatever can generate a set of different outcomes. The outcomes, or results derived from them, are then read and interpreted by the client or the diviner. The divining practices and their cultural embedding differ widely. They are roughly of two types. In one type, the emotional or spiritual state of the diviner during divination is of major importance; he (or she) becomes imbued with a heightened ability or power to comprehend the meaning of the result. The other type of divination, the type that interests us here, involves forms of divination that are shared, systematic, and structured approaches to knowledge. They depend not on the state of the diviner but on his careful adherence to procedures and on his reservoir of wisdom. These latter divination systems are, in fact, considered by some scholars to be sciences.

When we talk about using a randomizing mechanism, we mean, for example, spinning a roulette wheel, throwing dice, picking balls out of a jar, or flipping coins. We do not include, say, throwing a dart at a dartboard or spelling a word. The outcomes of the dart throw can change with skill and practice, and spelling depends on knowledge. In general, we are discussing randomizing mechanisms that have a fixed set of discrete outcomes, such as red, blue, and green; the integers

1–1000; or heads and tails. Even though we know the outcomes that may result, it is no simple matter for any of us to explain *why*, on a particular trial, one result occurred and not another. We might say that it was due to *chance* or *luck* or some such thing, but that would still need further clarification. There are occasions, however, when the involvement of outside forces is acknowledged. When, in the Old Testament, for example, Jonah is thrown overboard as the result of a casting of lots, it is believed that the outcome was guided or dictated by supernatural forces. It is not simply that Jonah was unlucky.

Belief in divination does not imply believing that *all* occurrences are controlled by extranormal forces; it is only believing that the outcomes of *particular* procedures, carried out under *particular* circumstances, and usually with *particular* materials, are expressions of specific deities, witches, or other supernatural forces. In classical Greece, for example, a form of divination called *astragalomancy* was practiced, in which the numbers 1, 3, 4, 6 were associated with the four sides of an astragalus (an ankle bone of a hoofed animal). The sides that faced upward when a set of five astragali was thrown were identified with a set of numbers, a god, and a prophecy. Similarly, among the Romans, four astragali were used. Because it was highly unlikely (using probability theory, we would calculate 1 chance in 10,000), the appearance of sides valued at 6, 6, 6, 6 was "The Throw of the Vulture" and foretold dire happenings. The throwing of these ankle bones has been linked to games of chance, the subsequent use of gaming dice, and even implicit and then explicit ideas about probability. The difference, however, is that classical Greek and Roman astragalomancy were parts of religious belief systems and were not games.

Although the procedures used in different forms of divination may begin with simple randomizing mechanisms, the outcomes can then be built upon in many and varied ways. We will discuss a few of these forms of divination, focusing in particular on the mathematical ideas that these procedures involve. We begin by looking briefly at knot divination in the Caroline Islands of the North Pacific, and then move to the practice of *Ifa* by the Yoruba, who live in and around Nigeria in West Africa. However, we will concentrate most fully on *sikidy*, the divination system used by the people who live on the large island of Madagascar.

2

The Caroline Islands are an archipelago that extends east to west for about 2400 km between the 5th and 10th parallels of north

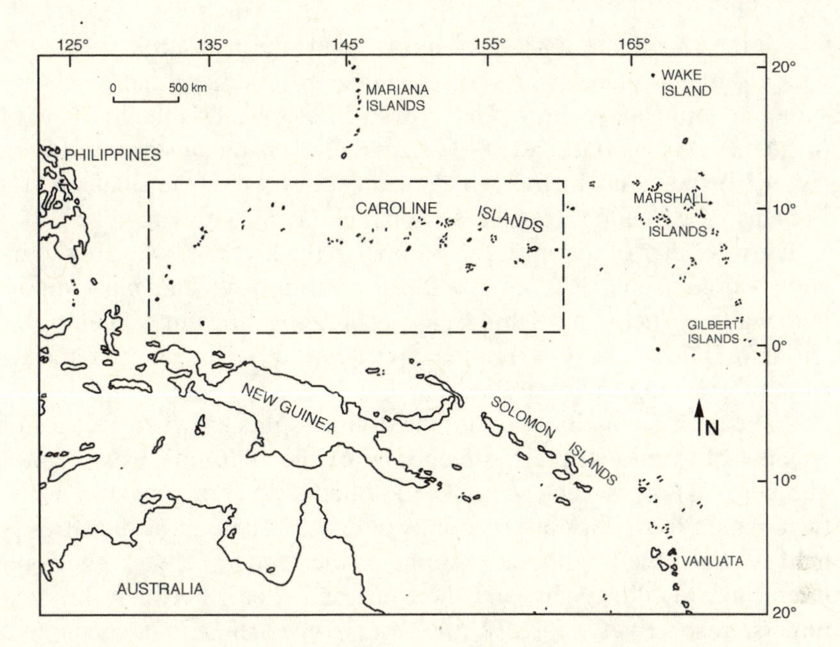

Map 1.1 The Caroline Islands in the North Pacific Ocean.

latitude. Between 1947 and 1986, the islands were part of the Trust Territory of the Pacific Islands administered by the United States. Now, they are the Federated States of Micronesia and Paulau. In all, the population is about 122,000 (see Map 1.1). Although the islands are spread out and separated by large distances across open waters, the people on the islands remain in close communication with each other. It is not surprising, therefore, that they share much in the way of culture, including a system of divination. There are variations from island to island, but there is substantial similarity. One myth about the origin of their divination system tells of a god, Supunemen, who brought to earth 16 destinies in the form of spirits. He had them build a sailing canoe in which they sailed from island to island teaching knot divination. When that job was accomplished, they went back to heaven never to return.

Each of the destiny spirits has a name, but what is more, each is associated with a pair of numbers. The first number in the pair can be 1, 2, 3, or 4, and the second number can be 1, 2, 3, or 4. Here, we represent the pairs as (a, b), where $a = 1, 2, 3, 4$ and $b = 1, 2, 3, 4$. The order in which the numbers appear is significant, that is (1, 3) is a different spirit than the pair (3, 1). In fact, in one version of the divination origin myth, where the spirits are both men and women, the men are the pairs in which a is less than b, while for the women, a is more than b. Corresponding pairs are spouses. Thus, for example, (1, 4) is the husband of (4, 1), (2, 3) is the husband of (3, 2), and (2, 4) is the husband of (4, 2). [As for the pairs in which in which a equals b, (4, 4) is the chief, and (2, 2) is his wife, while (3, 3) is their son, and (1, 1) is a young bachelor.]

The teachings of the spirits are known to the diviners, who are sacred and honored people. Not only are the diviners consulted on most important matters, including fishing, house building, traveling, naming of children, illness, and love, but they must carefully pass on their knowledge by teaching future diviners.

To begin a divination session, the diviner splits the young leaves of coconut trees into strips, and then they or the client makes a random number of knots in each strip. The knotted strips are placed in a pile from which four strips are randomly selected. The first of the strips is held between the thumb and forefinger, the second between the forefinger and middle finger, and the third and fourth between the next fingers, respectively. Finally, the knots on each strip are counted, returning, however, to a count of 1 each time a count of 4 is exceeded. That, in mathematics, can be described as counting *modulo 4*. If, for example, the strip had 9 knots, the final count would equal 1. [We would write it as $9 = 1 (\text{mod } 4)$.] A second strip of, say, 15 knots has a final count of 3, that is $15 = 3 (\text{mod } 4)$. Because beyond 4 the numbers recycle, the only results that can occur on each of these counts is 1, 2, 3, or 4. Thus, from these first two strips, a pair of counts (a, b) is obtained where $a = 1, 2, 3$, or 4 and $b = 1, 2, 3$, or 4. Each of the four possible values for a can be paired with each of the four values for b, and so the number of different pairs possible is the product $4 \times 4 = 16$. Hence, each pair (a, b) identifies one of the 16 destiny spirits.

Before any interpretation can be given, knots on the strips held between the next fingers are counted, also modulo 4. Again, the result is a pair of counts, (a', b'), identifying a second destiny spirit. The association and juxtaposition of the two destiny spirits are the basis for the diviner's interpretation. The coupling gives rise to a particular phrase or set of key words. Here, too, the order in which the results were obtained is significant. That is, just as (1, 3) is a different spirit than (3, 1), the pair of pairs [(1, 3), (2, 3)] elicits a different phrase than does [(2, 3), (1, 3)]. With 16 destiny spirits possible for the first pair, and 16 possible from the second pair, there are, in all, $16 \times 16 = 256$ different couplings that can result. The key words and phrases determined by the particular outcome are applied to the case at hand, taking into consideration the question, the questioner, and the surrounding circumstances. In all cases, the purpose of the divination is to gain information and understanding about ongoing or future happenings.

On one of the islands, in order to learn something about events that are in the remote future, a simpler divining scheme is used. The simplification reduces the number of possible outcomes from 256 to 16. As before, two pairs of count pairs, (a, b) and (a', b'), are formed. Before interpreting them, however, they are merged by the diviner into one pair (a'', b'') as follows:

$$a'' = (a+b) \bmod 4; \; b'' = (a'+b') \bmod 4.$$

In this method, the pairings in our example above [(1, 3), (2, 3)] would become the single pair (4, 1). Because the addition is modulo 4, once again the only possible values resulting for a'' and for b'' are 1, 2, 3, or 4, and so the final pair identifies one of the destiny spirits. In this simplified version, the scope of interpretation is limited to whether the far distant events will have favorable or unfavorable consequences.

3 Curiously, the numbers 16 and 256 are also prominent in Ifa divination among the Yoruba. However, the mode of divination and its goals, meaning, and surrounding belief system are decidedly different, as are the people, their environment, and their religion. Ifa is an especially significant form of divination because it has spread beyond its origins among the more than 18 million Yoruba in Nigeria to such groups as the Benin Edo of the same region, the Fon who established the Dahomey kingdom in the early 18th century in what is now Benin, the Ewe of Togo and Ghana, and the people of Cuba and Brazil who are descendants of Yoruba slaves (see Map 1.2). It has different names in different places as well as there being some variation in practices. While we look only at Ifa as it is practiced among the Yoruba of Nigeria, we recognize its greater spread and popularity.

Central to the beliefs surrounding the divination practice known as Ifa are: Ọlọrun, a deity who assigns and controls the destinies of mankind; Ifa, the God of Divination who interprets and transmits the wishes of Ọlọrun; and Eshu, the Trickster/Messenger of Ọlọrun who carries the offerings from the people to Ọlọrun, and who helps those who make appropriate offerings. The Yoruba deities are not limited to these; there are perhaps as many as 400 or 500 others.

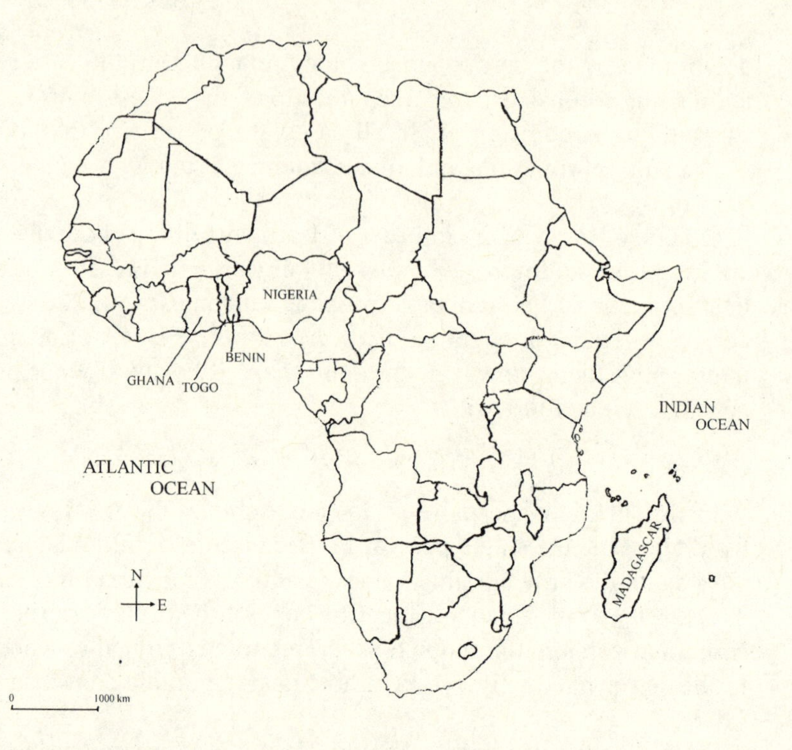

Map 1.2 Africa.

The Ifa divination procedure is a shared undertaking of the client and the diviner. The diviner knows the manipulations to be done and how they are to be read. He also knows–and this is what elevates him to a man of wisdom–a considerable number of verses to be said in response to the different outcomes. The verses contain the values, myths, mores, traditions, and theology of the Yoruba people. But also, specifically, each verse can be interpreted as a prediction with an attendant offering. What is most crucial is that, for each outcome, the diviner presents all of the verses he knows that are associated with it. The client, however, is the one who selects from these verses the one appropriate to his concern. In fact, if the client wishes, the diviner need not even be told the concern. The more verses a diviner knows for each outcome, the better the chances of one closely fitting the client's need. It is said that for a learner to be approved as a diviner by the senior diviners, he must know at least four verses for each outcome. Some diviners are said to know as many as 80 for some of them, but the diviners continue to learn throughout their years.

After attracting Ifa's attention by tapping sticks or bells, the diviner recites the necessary initial prayers and invocations. Then begins the casting of the outcomes. The procedure can utilize a set of 16 palm nuts, or a divining chain. We begin by describing the use of palm nuts, that is the nuts from inside the fruit of the *Elaeis guineesis idolatrica* or *King Palm*. The diviner beats the nuts together with both hands and then grasps a handful in his right hand, leaving only one or two nuts in his left hand. (If there are more than one or two remaining, the trial does not count and is ignored.) If *one* nut remains in the diviner's left hand, *two* short lines are made in the dust on his wooden tray; if *two* nuts remain, *one* short line is made on the tray. When questioned by a visitor as to why the number of marks and the number of remaining nuts were reversed, the diviners responded that this is the way they were taught by Ifa. I interpret this as implying that, to the diviners, the markings are not counts of the nuts remaining, as was assumed by the visitor. They are, instead, two different arbitrary symbols (|| and |) being used to distinguish between different occurrences. In all, this procedure is repeated until there are eight successful trials.

The marks on the tray are arranged into two columns of four each by alternating between the right and left columns on each trial. That is, the order in which the column positions are filled is as shown in Figure 1.1. Each position contains either one mark or two marks. If the nuts remaining on the eight trials were 1, 2, 2, 1, 1, 1, 2, 1, the resulting columns would be as shown in Figure 1.2. Each column results in one of the named Ifa figures. Because each of the four positions in a column can be filled in one of two possible ways, there are $2 \cdot 2 \cdot 2 \cdot 2 = 16$ different named Ifa figures that possibly could result. The meanings of the figure names are unknown, although in some myths the figures are described as the sons of Ifa.

2	1
4	3
6	5
8	7

Figure 1.1 The order in which column positions are filled when casting with palm nuts.

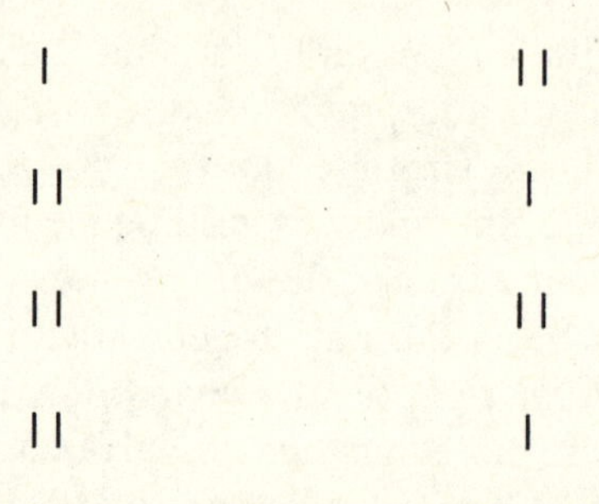

Figure 1.2 The outcome of a casting with palm nuts. The columns shown result from eight trials in which there remained 1, 2, 2, 1, 1, 1, 2, 1 nuts, respectively.

In our example (Figure 1.2), the name of the figure on the right is *Ofun,* and the figure on the left is *Ọbara.* The pairing, in this case *Ofun Ọbara* (readings are right to left), determines the verses to be recited by the diviner. As with the Caroline Island divination, the number of possible pairings is $16 \times 16 = 256$. And, in both systems, each of the 256 possible pairings seems equally likely to occur. But, quite different from the Caroline Islands, each Ifa outcome elicits a collection of verses rather than just a single phrase or set of key words, and, as was already noted, the Yoruba client plays a much larger role in the interpretation phase.

If, instead of palm nuts, the divining chain is used, a pair of columns would still result, with four positions each, to be associated with the named Ifa figures. The symbols for the figures, however, are slightly different. A divining chain contains eight halves of seed pods or seed shells (often the pods from the *ọpẹlẹ* or *Schrebera golungensis* tree) spaced out along the length of the chain such that they form two sets of four each, with a large space between them. The pod halves have two distinct sides: one showing the inside of the pod and the other showing the outside.

The pods are connected to the chain so that each can swing around freely and fall with either its inside or outside showing. As the diviner casts the chain onto a flat surface, he holds it in the space between the sets of pods so that the two segments of chain fall into parallel lines. Now the indicators are the insides and outsides of seed pods rather than the single or double marks made on a tray (see Figure 1.3). Again, there are two columns of four positions each, where each position can be filled in one of two ways. By associating a pod showing its inside with one mark (|) and a pod showing its outside with two marks (||), Figure 1.3 represents the same ordered pair, Ofun Ọbara, as did Figure 1.2. It calls forth the same verses from the diviner. An important difference,

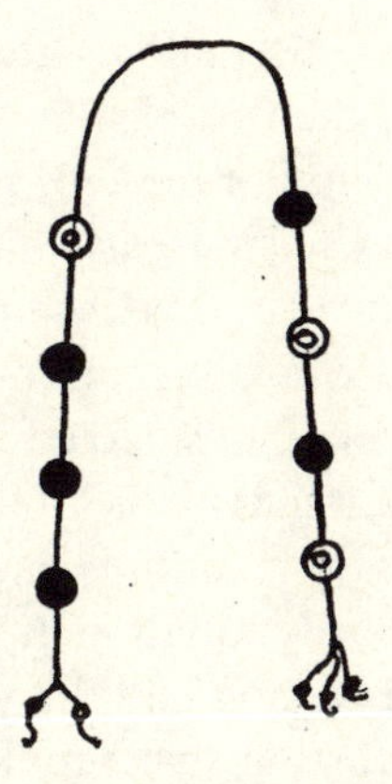

Figure 1.3 The outcome of a casting with a divining chain. ● indicates the outside of a pod; ◎ indicates the inside of a pod. (Notice that the ends are marked differently to insure that the same side is always made to fall on the right.) Compare this with Figure 1.2.

however, is that just one cast of the divining chain accomplishes the same end as eight successful trials using the palm nuts.

There is another aspect of Ifa divination that adds much to its richness and logical complexity. Once the chain or nuts are cast, but prior to the verses being recited, the client can ask a series of questions so phrased that the answers are choices between two specific alternatives, or among three alternatives, four, or even five. These questions and answers shed light on the issue at hand and help the client to select and interpret the verse he eventually chooses. The client can ask, for example, if the prediction will be for good fortune or for bad fortune. An answer of good fortune can be followed by asking which of the five desirable things in the world it will relate to: long life, money, wives, children, or victory over one's enemies. Juxtaposition of the outcomes of a series of castings determine which of the alternatives is the answer. The 256 figure pairs, that is, the possible outcomes of a single casting, have a specific rank order. The order is built upon certain principles. For example, pairs with repeated figures, such as Ọbara Ọbara, rank above pairs made up of different figures, and pairs containing the figure Ofun in either position have the same rank as those with Ọworin in that position. (Some rationales for the ranking involve seniority among the sons of Ifa, and, echoing the importance of twinness in Yoruba culture, equality of the ranking of Ofun and Ọworin involves an ongoing fight for position between two sons who are twins.)

For choosing between just two alternatives, the first stated alternative is the answer if the outcome of the first cast ranks higher than or equal to the outcome of the second cast. For three or more alternatives, there are nine specific outcomes on the first cast that stop the casting and indicate that the first alternative is to be chosen. If the outcome is not one of those nine figure pairs, the casting continues, and the alternative with the highest ranking outcome is selected as the answer. Again, equality of rank favors the alternative for which the figure pair first appeared. For this phase of the divination, because of its greater speed in obtaining each outcome, the divining chain is usually used. In the end, however, it is the original casting and the verse selected from the set of verses elicited by that casting that provide the prediction and, equally important, stipulate the necessary offering that is to be made. All of the other questions only elaborate and elucidate the prediction and offering.

4 Both modes of divination discussed so far, Caroline Island knot divination and Yoruba Ifa, involve randomizing processes and formal procedures. In the knot divination, the mathematical ideas are primarily numerical: counting modulo 4; addition modulo 4; identifying destiny spirits by ordered pairs of numbers; and linking significant words or phrases to ordered pairs of ordered pairs of numbers. By contrast, in Ifa, the mathematical ideas are primarily logical; that is, creating symbolic representations of the outcomes; using ordered pairs of these representations to elicit the verses, and comparing and selecting among ranked symbolic representations. *Sikidy*, the system of divination that is our next focus, has a long and broad history. Its practice is ongoing and is of great significance in Madagascar. It, too, starts with a randomizing process, and, as we shall see, the randomizing process has similarities to both the Caroline Island mode of knot counting and the Yoruba mode of representation of nut remainders. However, in Malagasy sikidy, as the divination continues, it involves mathematical ideas that are far more extensive.

Between the initial random process and the interpretation phase of sikidy, a multistep algebraic algorithm is followed. What is more, after applying the algorithm, the diviners use methods relying on the logical structure of the results to check that the algorithm has been properly carried out, and the interest and concerns of the diviners extend beyond the divination procedure to the symbolic forms themselves.

5 Madagascar is an island about 380 km off the east coast of Africa just opposite Mozambique (see Map 1.2). One of the largest islands in the world, it is approximately 1500 km long and 500 km wide and, in 1996, had a population of about 15 million people. Madagascar has about 20 different ethnic groups, but all of them share essentially the same language and much in the way of culture. The language is classed as a Western Indonesian subgroup of the Malayo-Polynesian language family, and so the earliest immigrants are thought to be originally from the Malayan–Indonesian archipelago. The numeral words, in particular, are considered almost identical to some others in that language family, and some groups on the island have outrigger canoes, quadrilateral houses on stilts, and circular fishing nets, all usually associated with Indonesian and Malayo-Polynesian culture.

Although its mechanism is unclear, there was, in addition, significant Islamic influence from about 750 CE to 1150 CE. While the religion itself was not adopted, traces of Arabic influence remain in the language, in particular in the names of months and of days, and in the use of Arabic script. At about the same time, there began substantial and continuous importation of African slaves and, hence, aspects of African cultures. This diversity of people and cultures was fused and politically organized through an expansionist, feudal-type monarchy established by the indigenous Sakalava people in the 16th century and then superseded in the early 19th century by the indigenous Merina kingdom. In 1896, the island was colonized by the French. Then, in 1959, Madagascar became the independent Malagasy Republic.

Throughout all the political and cultural changes of the past four centuries and despite internal political and cultural differences, some form of divination has remained in every part of Madagascar. There are variations, but there are basic essential similarities. And these similarities are intertwined with other shared aspects of the culture in which the divination is embedded, such as beliefs and practices related to ancestors and family tombs, residence and inheritance rules, and witchcraft. Because of the similarities, we will discuss Malagasy divination without distinguishing among the ethnic groups.

The *ombiasy*–a diviner who is expert in sikidy–specializes in guiding people. He has a long apprenticeship, a formal initiation, knowledge of formal divining practices, and, above all, an interactive approach in which the use and interpretation of the divining materials are combined with asking the client questions and then phrasing new questions to guide the next stage in the divination. The diviner discusses the interpretations with the client until the client determines the specific actions or answers that are relevant to solving his/her problems.

The important place of divination and the diviner in Malagasy culture can be seen from the broad array of questions brought for resolution. Some questions involve the day on which something should be undertaken whether it be a trip, planting, or ceremonial moving of the family tomb. There is a long tradition of adoption and fosterage, that is, placing children temporarily or permanently with other families. Hence, upon the birth of a child, the diviner is consulted to see how well the destinies of the child and its parents match or whether another family is preferable.

Other significant problems for resolution by divination are finding a spouse, finding lost objects, identifying thieves, and identifying the causes of illness, sterility, or any other misfortune. In Western medicine, for example, a virus may be considered the "cause" of an illness, but that does not answer the question of how, specifically, the illness was acquired, and why it was acquired by *that* individual at *that* time and in *that* place. To answer questions of cause, the Malagasy delve deeply, and the answers may well involve the actions of ancestors and/or witchcraft. Based on their knowledge and experience, some ombiasy are considered specialists and concentrate on dealing only with divination questions in their area of expertise.

As part of the initiation process of an ombiasy, the initiate must ceremonially gather for his subsequent use between 124 and 200 dried seeds of a fano tree (*Piptaenia chrysostachys*). To begin a divination session, the ombiasy, using various incantations, awakens the seeds in his bag and the verbal powers within him. The incantations include the origin myth of sikidy, which links it both to the return by walking on water of Arab ancestors who had intermarried with Malagasy but then left, and to the names of the days of the week.

The diviner takes a fistful of seeds from his bag and randomly lumps them into four piles. Each pile is reduced by deleting two seeds at a time until either one or two seeds are left in the pile. The four remainders become the entries in the first column of a tableau. If, for example, the four piles began with 21, 16, 19, 13 seeds, respectively, the remainders would be 1, 2, 1, 1 seeds. In effect, that is counting modulo 2: $21 (\text{mod } 2) = 1$, $16 (\text{mod } 2) = 2$, $19 (\text{mod } 2) = 1$, $13 (\text{mod } 2) = 1$. Since each of the four positions in the column can have one of two entries (either one seed or two seeds), the number of different columns that can result is $2 \cdot 2 \cdot 2 \cdot 2 = 16$. These 16 different possible outcomes are shown in Figure 1.4.

o	oo	o	oo	o	oo	o	oo
o	o	oo	oo	o	o	oo	oo
o	o	o	o	oo	oo	oo	oo
o	o	o	o	o	o	o	o

o	oo	o	oo	o	oo	o	oo
o	o	oo	oo	o	o	oo	oo
o	o	o	o	oo	oo	oo	oo
oo	oo	oo	oo	oo	oo	oo	oo

Figure 1.4 The 16 possible columns. Each column has four positions each of which can contain o (one seed) or oo (two seeds).

The entire process, beginning with another selection of a fistful of seeds from the bag, is then repeated three more times. Each time the results are placed in a column *to the left* of the previous column. Thus, the overall array is made up of four randomly generated columns of four entries each. With 16 different ways that each of the four columns can be filled, the total number of different possible arrays is $16 \cdot 16 \cdot 16 \cdot 16 = 65{,}536$.

This randomly generated set of data is called *the mother-sikidy*. An example is shown in Figure 1.5. To enable us to refer to the columns and resultant rows, we will call them, as is also shown in Figure 1.5, C_1, C_2, …, C_8 where C_1 through C_4 are the columns, and C_5 through C_8 are the rows.

The mathematical ideas involved in the algebraic algorithm that is next applied to the mother-sikidy fall within what modern mathemati-

C_4 ↓	C_3 ↓	C_2 ↓	C_1 ↓	
oo	o	oo	o	← C_5
o	o	oo	oo	← C_6
o	oo	o	o	← C_7
oo	o	o	o	← C_8

Figure 1.5 An example of a mother-sikidy. There are four randomly generated columns of four entries each where each entry is one seed or two seeds. Our labels for the columns, C_1, C_2, C_3, C_4, reflect the order in which they were placed. The same data, viewed horizontally, will be referred to as C_5, C_6, C_7, C_8.

cians refer to as *Boolean algebra* and *two-valued logic*. Before proceeding to the specific algorithm, let us draw together some of *our* relevant mathematical ideas.

6 In 1937, in his master's thesis at MIT, Claude E. Shannon introduced the use of symbolic logic to simplify switching circuits. One of his examples was the automatic addition of binary (base two) numbers using only relays and switches. Base two numbers involve only the two different symbols 0, 1 (called *bits*, a contraction of binary digits), as contrasted to base ten numbers in which there are the ten different digits 0, 1, 2, ..., 9. Also, in a base ten multidigit number, each consecutive position to the left is worth one higher multiple of ten, while in a base two number, the consecutive bit positions are worth one higher multiple of two.

Example

Base ten number : $4035_{10} = 4\,(1000) + 0\,(100) + 3\,(10) + 5.$

Base two number : $1011_2 = 1\,(8) + 0\,(4) + 1\,(2) + 1.$

The application introduced by Shannon was new and of extreme importance as base two number representation and its automatic electrical manipulation became the internal operating mode for digital computers. The algebra of logic that he used, however, was about 100 years old, having been developed primarily by George Boole and discussed in his 1847 *The Mathematical Analysis of Logic.* One of Boole's stated concerns was to express "logical propositions by symbols, the laws of whose combinations should be founded upon the laws of the mental processes which they represent..." Further, the symbols need not be interpreted as magnitudes but "every system of interpretation...is equally admissible, and it is thus that the same process may, under one scheme of interpretation, represent the solution of a question on the properties of numbers, under another, that of a geometrical problem, and under a third, that of a problem of dynamics or optics." Or, as we are adding here, the interpretation may represent the solution of a problem in divination.

In Boole's system, there are, first of all, symbols that represent classes of objects, then rules of operation on the symbols, and, finally, the observation that the rules he established are the same as would hold in the two-valued numerical algebra of 0 and 1. Boole's basic operations on classes x and y were forming a new class of things that are either x's *or* y's but not both (referred to here as OR and symbolized by "+"), and forming a new class by selecting things that are both x's *and* y's (referred to here as AND and symbolized by "·"). In terms of 0 and 1, the results of these operations are:

AND (·)	OR (+)
$0{\cdot}0 = 0$	$0 + 0 =$ no interpretation
$1{\cdot}1 = 1$	$1 + 1 =$ no interpretation
$1{\cdot}0 = 0$	$1 + 0 = 1$
$0{\cdot}1 = 0$	$0 + 1 = 1$

Later, Jevons, in his book *Pure Logic* of 1864, modified the OR operation, and, in what is *now* known as Boolean algebra, it is:

OR (+)
$0 + 0 = 0$
$1 + 1 = 1$
$1 + 0 = 1$
$0 + 1 = 1$

Notice that, in general, that is, whether $x = 0$ or $x = 1$, $x{\cdot}x = x$ and $x + x = x$. Both the OR and AND operations are what we call *commutative* and *associative*. To be commutative means that the result remains the same regardless of the order of the elements to which the operation is applied. That is, $x{\cdot}y = y{\cdot}x$ and $x + y = y + x$. (A more familiar example of commutivity is our ordinary addition where $a + b = b + a$. Our ordinary subtraction, however, is not commutative, that is, $a - b \neq b - a$.) Associativity refers to the grouping of operations. It means that consecutively applied operations give the same result regardless of which is carried out first. That is, $x{\cdot}(y{\cdot}z) = (x{\cdot}y){\cdot}z$ and $x + (y + z) = (x + y) + z$. [Again, using our ordinary arithmetic as an example, multiplication is associative, $a(bc) = (ab)c$, but division is not, $a/(a/c) \neq (a/b)/c$.] For the two-valued case of commutivity, observe that $0{\cdot}1 = 1{\cdot}0 = 0$ and $0 + 1 = 1 + 0 = 1$. For associativity, there are many more cases to verify, for example, $0 + (1 + 1) = (0 + 1) + 1 = 0$. In Boolean algebra, in addition to the OR and AND operations, there is also an operation creating Not-x, now more commonly known as $\bar{x}$ or as taking the *complement* of x. For the two-valued case, when $x = 0$, $\bar{x} = 1$ and for $x = 1$, $\bar{x} = 0$.

For Shannon, concerned with the construction of circuits, the OR operation was seen as analogous to placing switches in parallel and the AND operation to placing them in series. He further identified the values 1 and 0 with the states of the switches being closed and open, respectively, and hence with the transmission or non-transmission of electrical pulses. Thus, for two switches in series, both must be closed to transmit a pulse ($1 \cdot 1 = 1$), while if one or the other or both are open, there is no pulse transmitted ($0 \cdot 1 = 0$, $1 \cdot 0 = 0$, $0 \cdot 0 = 0$). This case is familiar in strings of decorative lights, such as those used on Christmas trees. When there are two or more lights on a string, if one light is not operating (hence, not transmitting a pulse), the overall string does not operate. However, for two switches in parallel, at least one must be closed for the circuit to transmit a pulse ($1 + 1 = 1$, $1 + 0 = 1$, $0 + 1 = 1$), while only if both are open is no pulse transmitted ($0 + 0 = 0$). Here, a common situation is, say, a radio and a toaster plugged into a double outlet in your kitchen. Depressing the toaster button–closing its switch–starts the toast whether or not the radio is playing, and, similarly, the radio works with or without the toaster. Only if neither is in use does no electricity flow. In electrical circuits, an inverter is analogous to the Boolean complement, that is, it changes a pulse to no pulse and vice versa ($\bar{1} = 0$, $\bar{0} = 1$).

Shannon's goal was to construct circuits that could carry out arithmetic processes. This is greatly simplified by using numbers in base two form, where only the two symbols 0 and 1 are involved in the arithmetic. His accomplishment was to combine the basic operations of the two-valued Boolean algebra, so that the result would be the sum of the binary numbers, and then to use his switching analogies to physically construct such a circuit. By first forming a Boolean expression for the circuit, other logically equivalent expressions could be derived from it algebraically; from among them, the simplest could be selected for actual physical construction.

Combinations of the basic AND, OR, and complement operations can be considered Boolean operations in and of themselves. The operation of particular interest here because of its central role in sikidy is the *exclusive or*, referred to in current computer science texts as XOR and denoted by "$\oplus$". The circuit for this operation plays a significant role in computer circuitry as it is an important component of Shannon's binary adder. For 0 and 1, the results of an XOR operation are:

XOR ($\oplus$)
$0 \oplus 0 = 0$
$1 \oplus 1 = 0$
$1 \oplus 0 = 1$
$0 \oplus 1 = 1$

This operation is also commutative and associative; that is, $x \oplus y = y \oplus x$ and $(x \oplus y) \oplus z = x \oplus (y \oplus z)$.

Further, it has the properties that, whether $x = 0$ or $x = 1$, $x \oplus x = 0$ and $x \oplus 0 = x$. (In formal mathematical terminology, 0 and 1 form an Abelian group under XOR, with 0 the identity element and each element its own inverse.) In the symbolism of Boolean algebra, $x \oplus y$ can be expressed as $\bar{x} \cdot y + x \cdot \bar{y}$, or in numerous other ways including $(x + y) \cdot (\overline{x \cdot y})$. The switching circuit embodiment for the latter is shown in Figure 1.6.

As well as being a component of a computer's binary adder, the current importance of XOR also stems from its use in the detection of errors in the electronic transmission of binary data. The procedure involved is referred to as *even parity checking*. Received binary data are checked to see whether the total number of 1s in it is odd or even. To insure that it will be even, an additional appropriate bit is sent along with the data. Thus, if the checking yields an odd number of 1s, there was an error in the transmission. The XOR easily handles this parity checking: combining any number of 0s and 1s consecutively via XOR gives a 1 if the number of 1s is odd and an 0 if the number is even. We find, in sikidy, that in addition to XOR being the central operation, the mode of even parity checking is also used.

In sikidy, using the modern algebraic terminology, the symbols for the classes in its two-valued logic are o (one seed) and oo (two seeds). For these, the results of the XOR operation are:

XOR ($\oplus$)
o $\oplus$ o = oo
oo $\oplus$ oo = oo
o $\oplus$ oo = o
oo $\oplus$ o = o.

(Although the ombiasy surely did not think of this as *the exclusive or* and did not call the operation he applied XOR, he did use these rules of combination. We, therefore, will use our descriptor XOR when describ-

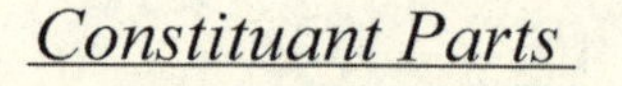

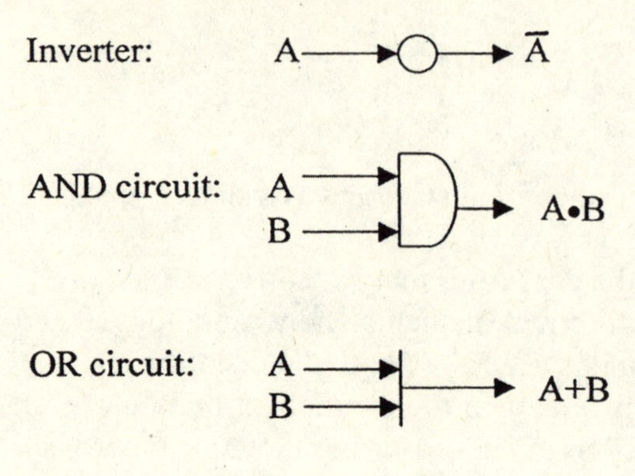

$x \oplus y = (x+y) \bullet (\overline{x \bullet y})$

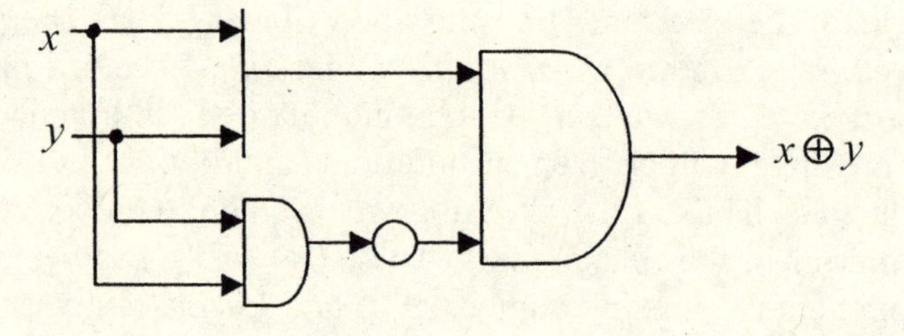

Figure 1.6 XOR circuit diagram.

ing what he did.) The commutative and associative properties of XOR, of course, remain unchanged. Now, however, oo is the identity; that is, $x \oplus$ oo $= x$ and $x \oplus x =$ oo. Another visualization of XOR that may assist in following the sikidy algorithm and checking procedures is to associate o (one seed) with odd and oo (two seeds) with even. Then, the rules of combination of XOR behave like the addition of odd and even numbers:

$$\begin{aligned} \text{odd} \oplus \text{odd} &= \text{even} \\ \text{even} \oplus \text{even} &= \text{even} \\ \text{odd} \oplus \text{even} &= \text{odd} \\ \text{even} \oplus \text{odd} &= \text{odd} \end{aligned}$$

with *even* acting as the identity; that is, for $x =$ odd or even, $x \oplus$ even$=x$ and $x \oplus x =$ even.

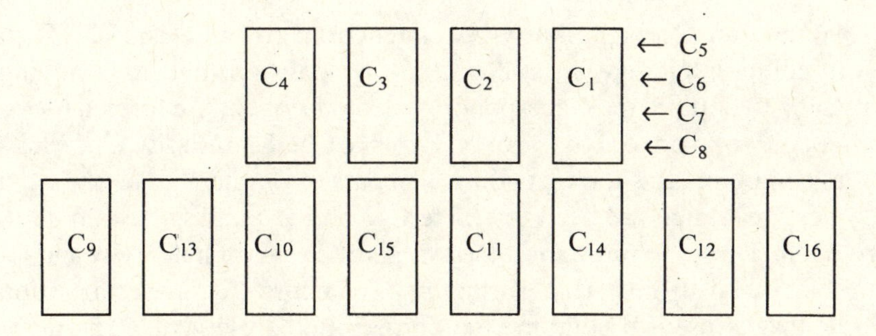

Figure 1.7 The placement of the columns.

7 With these ideas in hand, we return to the procedures of sikidy. There was, so far, the mother-sikidy, a randomly generated set of data whose four columns and four rows we called C_1, C_2, C_3, C_4 and C_5, C_6, C_7, C_8, respectively (an example was Figure 1.5).

Next, using the XOR operation and these columns and rows, an additional eight columns are generated. The algorithm for their generation involves not only what is to be calculated but the order in which it is to be done and where each result is to be placed. Just as we used C_1, C_2, C_3, C_4 to reflect the order in which the initial random columns were generated, we will use C_9, C_{10}, …, C_{16} to reflect the order in which the next eight columns are generated. Figure 1.7 shows where each of them is placed.

The algorithm begins with the formation of column 9 based on the contents of rows 8 and 7: $C_9 = C_8 \oplus C_7$, that is, position by position, moving from right to left, the elements of C_8 and C_7 are combined via XOR to produce the corresponding elements of C_9. For the example of Figure 1.5 where the elements of C_8 and C_7 are

o   oo   o   o ← C_7

oo   o   o   o ← C_8

the calculation would be o $\oplus$ o = oo, o $\oplus$ o = oo, o $\oplus$ oo = o, and oo $\oplus$ o = o, with the result that C_9 is

oo

oo

o

o

Next, column C_{10} is similarly generated from rows C_6 and C_5. Then from columns C_4 and C_3, still position by position, but now moving from top to bottom, the corresponding elements of C_{11} are formed (oo ⊕ o = o; o ⊕ o = oo; o ⊕ oo = o; oo ⊕ o = o), and, similarly, C_{12} results from combining the corresponding elements of C_2 and C_1. Next, C_9 and C_{10} just generated are used to create C_{13}, and C_{11} and C_{12} combine to give C_{14}. These, in turn, are joined to give C_{15}, and finally, C_{15} and C_1 are combined to give C_{16}. Throughout, of course, it is the operation XOR that is used to combine the corresponding elements. The algorithm can be summarized as follows:

$$\begin{aligned} C_9 &= C_8 \oplus C_7 \\ C_{10} &= C_6 \oplus C_5 \\ C_{11} &= C_4 \oplus C_3 \\ C_{12} &= C_2 \oplus C_1 \\ \hline C_{13} &= C_9 \oplus C_{10} \\ C_{14} &= C_{11} \oplus C_{12} \\ \hline C_{15} &= C_{13} \oplus C_{14} \\ \hline C_{16} &= C_{15} \oplus C_1 \end{aligned}$$

Applying the algorithm to the example of the initial mother-sikidy in Figure 1.5, the final tableau is shown in Figure 1.8. It would be beneficial to follow the algorithm through the example. Notice, as you do, the spatial as well as algebraic aspects of the algorithm. For example, it is the adjacent columns or rows that are combined, but the results are so placed that *their* combination is then centered between them: C_{13} goes between C_9 and $C_{10;}$ C_{14} is between C_{11} and $C_{12;}$ and C_{13} and C_{14} are equally spaced to either side of C_{15}. Columns 9 through 16 are referred to by the Malagasy as the *descendants* of the mother-sikidy. Each column in the final tableau has a referent, such as, C_4 is associated with the earth, C_{11} with the ancestors, and C_{15} with the creator (the referents are in Table 1.1).

Before proceeding with the interpretation phase of the divination, the ombiasy uses his knowledge of the logical structure of the tableau to check that the algorithm has been carried out correctly. He knows several relationships that the tableau *must* contain regardless of the initial random data. These relationships are particularly interesting because of their differences: one involves examination of the overall tableau; another involves examining the results of combining some

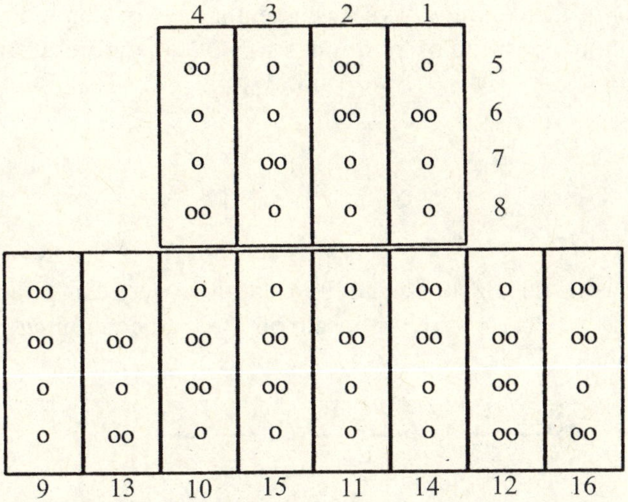

4	3	2	1	
oo	o	oo	o	5
o	o	oo	oo	6
o	oo	o	o	7
oo	o	o	o	8

oo	o	o	o	o	oo	o	oo
oo	oo	oo	oo	oo	oo	oo	oo
o	o	oo	oo	o	o	oo	o
o	oo	o	o	o	o	oo	oo
9	13	10	15	11	14	12	16

Figure 1.8 An example of a final tableau. (C_1, …, C_8 contain the same initial random data as is shown in the mother-sikidy in Figure 1.5. C_9, …, C_{16} are her descendants.)

particular columns; and yet another, the parity check, involves examining only one specific column.

First of all, the ombiasy knows that at least two of the 16 C_i must be the same. In order to assure *ourselves* of this, we will prove that it is true, using a method called a *proof by contradiction*. Consider that there are 16 C_i and there are 16 different possible columns that can have one or two seeds in each of its four positions. So, if all 16 C_i were different, they must contain one of each of the possible 16. Looking at the 16 possible columns in Figure 1.4, observe that in each of the four posi-

Table 1.1 Column referents

C_1: the client (the person seeking the consultation)
C_2: material goods
C_3: a male evil-doer (lit.–the third)
C_4: the earth
C_5: the child
C_6: the bad intentions
C_7: a woman
C_8: the enemy (lit.–the eight)
C_9: the spirit (lit.–the ninth)
C_{10}: nourishment
C_{11}: the ancestors
C_{12}: the road
C_{13}: the diviner
C_{14}: the people
C_{15}: the creator
C_{16}: the house

tions, eight of the columns have oo and eight have o. Combining all 16 of the columns, position by position, via XOR, therefore, would give

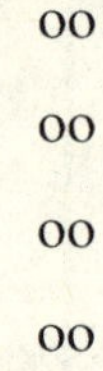

However, using the algebraic definitions of the descendant C_i and keeping in mind that $2(x) = x \oplus x = \text{oo}$ and $\text{oo} \oplus x = x$, combining all of the C_i in a tableau yields

$$\underbrace{C_1 \oplus C_2}_{C_{12}} \oplus \underbrace{C_3 \oplus C_4}_{C_{11}} \oplus \underbrace{C_5 \oplus C_6}_{C_{10}} \oplus \underbrace{C_7 \oplus C_8}_{C_9} \oplus C_9 \oplus C_{10} \oplus C_{11}$$

$$\oplus C_{12} \oplus \underbrace{C_{13} \oplus C_{14}}_{C_{15}} \oplus C_{15} \oplus C_{16}$$

$$= 2(C_9 \oplus C_{10} \oplus C_{11} \oplus C_{12} \oplus C_{15}) \oplus C_{16} = C_{16} = C_{15} \oplus C_1.$$

If this combination gave the same result as combining the 16 different possible outcomes, that is, if it were

oo

oo

oo

oo

it would necessarily mean that in each of the four positions, C_{15} and C_1 were either both o or both oo, and hence, C_{15} and C_1 are always both the same. *The assumption that the 16 C_i were all different has led to a conclusion that these two C_i are the same.* This contradiction tells us that the assumption must have been incorrect: the 16 C_i can *not* include one of each of the 16 possible results and so must include some repetition. In our example (Figure 1.8), due to the particular initial data, there are several repetitions (such as C_1 and C_{11} or C_3 and C_7) within the tableau, but, as we have shown, there must be some.

The next check carried out by the ombiasy involves what they call "the three inseparables." The inseparables are three particular pairs of C_i: C_{13} and C_{16}; C_{14} and C_1; and C_{11} and C_2. The ombiasy know that the results obtained when combining each pair via XOR is always equal to the results for the others. Again, using the algebraic definitions of the columns, we can prove to ourselves that this must be the case:

$$C_{13} \oplus C_{16} = C_{13} \oplus (C_{15} \oplus C_1) = C_{13} \oplus (C_{13} \oplus C_{14}) \oplus C_1$$

$$= C_{14} \oplus C_1 \text{ and}$$

$$C_{14} \oplus C_1 = (C_{11} \oplus C_{12}) \oplus C_1 = C_{11} \oplus (C_1 \oplus C_2) \oplus C_1 = C_{11} \oplus C_2.$$

In the example in Figure 1.8, these "inseparables" all, of course, give the same result, namely

o

oo

oo

oo

The next and final check is the parity check: the creator (C_{15}) must contain an even number of seeds. In the example in Figure 1.8, it surely does. To show this generally, however, we will show that when the four positions within C_{15} are combined via XOR, the result must be oo. From the algebraic definitions

$$C_{15} = C_{13} \oplus C_{14} = (C_9 \oplus C_{10}) \oplus (C_{11} \oplus C_{12})$$
$$= (C_1 \oplus C_2) \oplus (C_3 \oplus C_4) \oplus (C_5 \oplus C_6) \oplus (C_7 \oplus C_8).$$

Focusing on just the first element in C_{15}, this statement tells us that its value results from combining the four elements across the first row of the mother-sikidy and the four elements down the first column. Similarly, the second element in C_{15} comes from combining the elements across the second row and down the second column. The third element comes from the third row and the third column, and the fourth element from the fourth row and fourth column. Then, when these four elements of C_{15} are combined, it effectively utilizes each of the elements of the mother-sikidy *twice*–once from the contributions going across the rows and once from going down the columns. Since, for any x, $x \oplus x =$ oo, using each of the elements twice must result in oo.

Only if all the results are correct does the diviner continue with his complex and discussive interpretations. As the divination continues beyond the beginning standard algorithm, there are about 100 more formulas he can call into play, all built upon the same initial data

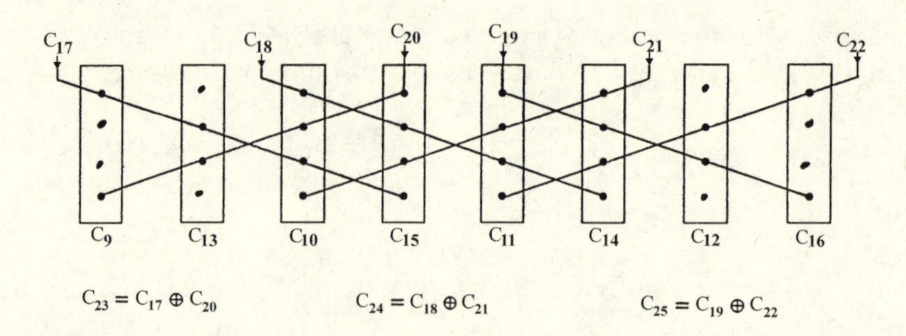

Figure 1.9 A secondary series using diagonal readings from the tableau. C_{17}, for example, is the first element of C_9, the second element of C_{13}, the third of C_{10}, and the fourth of C_{15}.

and using the same two-valued logic. For example, one secondary series is obtained by reading columns diagonally rather than up and down or right to left. Figure 1.9 shows how C_{17}, C_{18}, ..., C_{22} are read and then combined via XOR to give C_{23}, C_{24}, and C_{25}. In Figure 1.10 another, more intricate, reading pattern is used to define the series C_{26}, ..., C_{29} which, when combined, gives C_{30} and C_{31}.

In every case, no matter how they are read or combined, the result is always one of the 16 possible columns made up of four entries each. The diviner's interpretation of the results and the generation and interpretation of additional results depend on which of the 16 appear and on their juxtaposition to each other. Interpretation is where the logical algebra leaves off and the attribution of meaning begins. The interpretations are not rote or standardized and vary with the ethnic

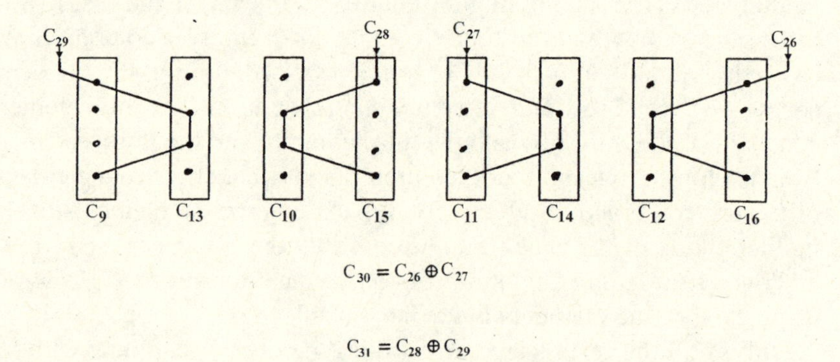

Figure 1.10 Another secondary series using a patterned selection of elements from the tableau. C_{26}, for example, is the first element of C_{16}, the second and third of C_{12}, and the fourth of C_{16}.

group, the diviner, and, above all, the situation under discussion and the course of the discussion. There are, however, some shared themes that we explore further because they involve additional mathematical ideas.

First of all, the diviners identify the 16 possible outcomes by names, although the names may differ in different regions. Some, but not all, of the names are related to the names of the months. Also, for many diviners, the 16 outcomes have particular directional associations. This imposition of spatial orientation on the outcomes echoes other aspects of Malagasy culture. In particular, there is strong belief in astrology and, in that, 12 month names are associated with 12 radial directions. Further, directionality is a significant part of daily life. A north–south axis and an east–west axis determine the positioning of houses in a village and the positioning of rooms within a house. Ordering along the north–south axis reflects relationships among the living, and ordering along the east–west axis reflects relationships between the dead and living. Thus, for example, houses of the village founder are in the northeast corner of a village, and, if possible, a son builds to the southwest of his father's house. The north and east are associated with men, adults, seniors, kinsmen, the dead, and royalty, in contrasted to the south's and west's association with women, children, juniors, strangers, the living, and commoners. Also, different directions have different values–the northeast is good, while the southwest is lacking in virtue, and the directions between vary in religious and moral value. As a result, there are not only prescribed house and tomb orientations, but also prescribed interior layouts extending to where specific items need be stored and how visitors position themselves when gathered in a room. The positioning of visitors joins direction with social status–the most important stand toward the northeast and the least important toward the southwest.

The Malagasy concern for spatial positioning and its social concomitants are reiterated in sikidy. Each of the 16 possible columns is conceived of by the diviners as associated with a particular place in a square. The axes of the square are oriented north–south and east–west with each of the sides, therefore, identified with one of the four cardinal directions. The eight possible outcomes that contain an even number of seeds are categorized as *princes*, and the eight with an odd number of seeds are *slaves*. Their positions in the square are shown in Figure 1.11.

The square is separated into halves by the diagonal that joins its northeastern and southwestern corners–the northwestern half is called the Land of Slaves, and the southeastern half is called the Land of

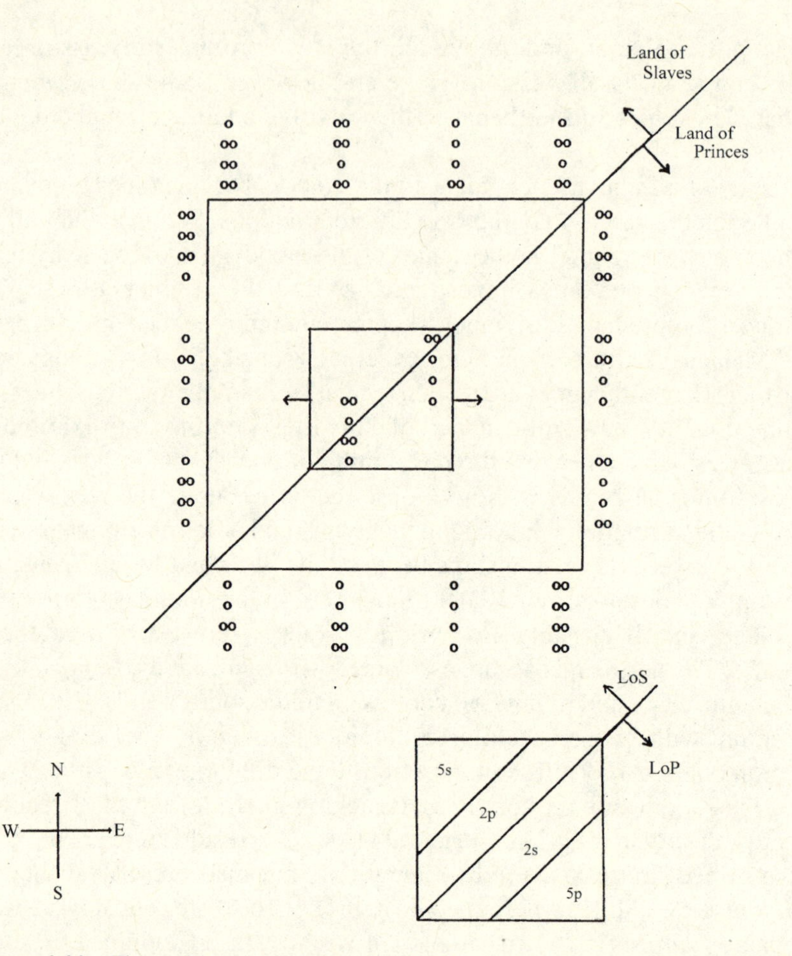

Figure 1.11 The positions associated with the 16 possible outcomes. The two in the center are the migrators. (Note the symmetric placement of slaves and princes in their respective lands.)

Princes. However, the Land of Slaves does not only contain slaves, nor does the Land of Princes only contain princes. As can be seen in Figure 1.11, there are slaves and princes in each half, but there are also two migrators, that is, one slave and one prince that move, more or less, with the sun, and so their place depends on the time of day of the divination. From sunup to 10 a.m. the migrators are in the east, from 10 a.m. to 3 p.m. in the north, from 3 p.m. to sunset in the west, and never in the south as divination does not take place at night. (Some diviners always associate the migrators with the west.) Power inequalities result from rank and place: princes are more powerful than slaves;

slaves (or princes) from the Land of Princes are more powerful than slaves (or princes) from the Land of Slaves; slaves from the same land are never harmful to each other; and battles between two princes from the Land of Princes are always serious but never end in death.

An example of a divination that makes use of these relationships is one related to illness. If, in the final tableau, the client (C_1) and the creator (C_{15}) are the same, there definitely will be recovery; if the client and the ancestors (C_{11}) are the same, the illness is due to some discontent on the part of the ancestors; and if the client and the house (C_{16}) are the same, the illness is the same as an earlier illness from which there has been recovery. The result of the combination $C_1 \oplus C_9$ has the illness itself as a referent. If, for example, the client is a slave of the east, and the illness is a prince of the south, the client is dominated by the illness, and so it is serious. However, since both the east and south are in the Land of Princes, the illness will not lead to death. If, however, the client were a prince of the north (Land of Slaves), there would be a strong battle with a good chance that the ill person would die. Some tableaux are considered to be exceptionally serious and quite hopeless. The most extreme is "the red sikidy", in which C_1, C_2 and C_3, and C_4 are all

oo

oo

oo

oo

In this case, all 16 C_i in the tableau are the same. Some of the initial questions and answers in a divination related to illness are straightforward, but in general, the divination will continue with less programmatic combinations to answer further questions on causes and cures.

The attribution of directionality to the outcome gives rise to tableaux with special importance. The power to see into the past or future is greater if all four regions, east, west, north, and south, are represented in a tableau. Tableaux with the most power, however, are those in which all four regions are represented, but at least one region has only one representative. These distinctive tableaux are referred to as *sikidy-unique*, and they hold special abstract interest for the ombiasy. In Figure 1.12a, C_8 is the sole representative of the south. (Although C_{10} is a migrator that causes no problem as a migrator can never be in the south.) The tableau in Figure 1.12b is even more unusual in that it has the creator (C_{15}) as the only representative

(a)

4(E)	3(E)	2(W)	1(E)	
oo	oo	o	oo	5(E)
o	o	oo	oo	6(E)
oo	o	oo	o	7(N)
oo	oo	o	o	8(S)

oo	oo	oo	o	oo	o	o	o
o	oo	o	oo	oo	oo	oo	oo
o	oo	o	oo	o	oo	o	o
oo	o	o	o	oo	oo	oo	oo
9(E)	13(W)	10(Mig)	15(W)	11(N)	14(N)	12(N)	16(N)

(b)

4(S)	3(W)	2(S)	1(S)	
o	oo	o	o	5(S)
o	oo	o	o	6(S)
oo	oo	oo	oo	7(S)
o	o	o	o	8(S)

o	o	oo	oo	o	o	oo	o
o	o	oo	oo	o	o	oo	o
o	o	oo	o	oo	oo	oo	o
o	o	oo	o	oo	oo	oo	oo
9(S)	13(S)	10(S)	15(E)	11(S)	14(S)	12(S)	16(N)

Figure 1.12 Examples of sikidy-unique. The associated direction is shown for the specific outcome in each column. In (a), C_8 is the sole representative of the south. In (b), C_{15}, C_3, C_{16} are the sole representatives of the east, west, and north, respectively.

from the east, as well as having a single representative from the west (C_3), and a single representative from the north (C_{16})–all the rest of the C_i are from the south. The interest of the ombiasy in the sikidy-unique extends well beyond any practical or divinational needs. These special forms are sought by the ombiasy for themselves, that is, in addition to simply encountering them in the course of divinatory consultations, finding beginning data that lead to such tableaux is an intellectual pursuit in and of itself. Knowing as many as possible increases prestige: some such tableaux are publicized by being posted on doors; some are shared with other ombiasy by word of mouth; and there is speculation, but with no persuasive evidence, that some ombiasy have secret rules for generating certain types of sikidy-unique. No one knows all of them or how many there are, and so the search for them continues.

8 Most often, the way that mathematical ideas are expressed in a traditional culture is unique to that culture or, at most, is shared with nearby neighbors. For sikidy, however, this is not the case. There are several reasons to believe that sikidy was influenced by an early form of Arabic divination called *ilm er-raml* or *science of the sand*. First, as we noted before, Arab ancestors are described in the sikidy origin myth as told by the diviners at the start of a divinatory session. Also, Islamic month names are used within the Malagasy divination, and there are some writings about the divination in a Malagasy script derived from the Arabic. This early Arabic influence on Madagascar has been linked by historians to the Arabic sea-going trade in the ninth or 10th century CE, involving the southwest coast of India, the Persian Gulf, and the east coast of Africa.

Communication, however, goes in both directions. If Arabic ideas influenced Malagasy divination, it is quite likely that, in turn, Malagasy ideas influenced the Arabic version. Bernard Carra de Vaux, a historian of mathematics, wrote in the early 20th century about his investigation into the origin of the Arabic ilm er-raml. He cited Zénâti of the Berber subgroup, the Zénâtah, as a major author but further notes that students of Zénâti list earlier teachers of Zénâti extending back to a Berber contemporary of Mohammed and before him to Tomtom el-Hindi, and still earlier going back to Idrîs, the god of writing. In any case, according to other writers, the sand divination was spread by Arab scholars, probably in the eighth and ninth centuries, to Damascus, to Alexandria, to Cairo, into the Sudan, into Spain, and, later, into numerous other places including France and Germany. And there are a number of places where some forms of it still exist. Further, it has been argued that some of the ongoing varieties in West Africa predated, and fed into, the Arab version.

We know that in the 12th century, when an Arabic text was translated into Latin, ilm er-raml entered Europe under the name of *geomancy*.

There are other unrelated usages of the word, but as this particular divination mode, geomancy became widespread throughout Europe where it remained popular through the 16th century. Thus, we see the central algorithm and parity check of sikidy shared by numerous people, including the Malagasy, the Arabs, and the Europeans. The Arabs and Europeans, however, do not appear to share the Malagasy use of seeds, the additional checks used by the ombiasy, their additional combining formulas, or anything comparable to sikidy-unique. There is no way to know exactly where or by whom the idea was started. Clearly, it was shared by several peoples as they interacted, and, in the process of sharing, it was modified, mixed with other ideas, and adapted to different cultures.

Sikidy, then, is another branch of the tree-like history of ilm er-raml/ geomancy, and hence, some of the mathematical ideas in it are shared with others. Wherever the ideas came from, when integrated into sikidy, they became intertwined with, and tailored to, Malagasy culture. The ombiasy became fluent in handling the mathematical ideas included in the initial randomizing process, the two-valued logic, an extensive algorithm involving both algebraic and spatial patterning, checking procedures, and directional attributions of the outcomes. And we find especially noteworthy that the ombiasy went beyond the divination in their interest in these ideas. The search for special forms underscores the fact that theirs is an active intellectual engagement rather than just a complex, but rote, process.

Sikidy, ilm er-raml, and their European cousin, geomancy, make us keenly aware that *in the context of divination*, the mathematical ideas they contain were circulating among different peoples in different cultures for hundreds of years prior to what is considered to be their emergence in modern mathematics.

9 Divination, whether with as extensive a procedure as Malagasy sikidy, or more limited in its steps as Yoruba Ifa or Caroline Island knot divination, has an unusually broad spread. Because of the particular randomizing methods they contain, the forms that we examined here share 16 as a significant number. We encountered it arising in $4 \cdot 4 = 16$, $2 \cdot 2 \cdot 2 \cdot 2 = 16$, $16 \cdot 16 = 256$ and $16 \cdot 16 \cdot 16 \cdot 16 = 65{,}536$. In part, these result from making several binary choices. Many contemporary scholars interested in cognitive processes believe that the creation of dichotomies, such as light/dark, thick/thin, odd/even, on/off, yes/no, up/down, and hence, binary choices, is fundamental to all

human thinking. So, while often found in divination, they are not a necessary or defining property of divining. Knot divination, for example, did not involve a binary choice: the counting modulo 4 could just as well have been modulo 5. For modulo 5, when combining a consecutive pair of outcomes, the number of possible pairs would be $5 \cdot 5 = 25$, and the number of possible pairs of pairs would be $25 \cdot 25 = 625$. Similarly, a four-sided astragalus could just as well be a six-sided cubic die or a five-sided polyhedron.

In any case, divination begins with a discrete randomizing process that yields a finite number of possible outcomes. This initial process may be followed by other procedures. As we have seen, it is these additional procedures that vary considerably and range from brief to extended. What is crucial, and what ties the divination practices together, is that these additional procedures are systematic and clearly spelled out. It is, therefore, not surprising that many, if not most, of the various modes of divination have mathematical ideas as an integral component.

NOTES

1. Particularly recommended for general understanding of divination are the chapters by Philip Peek ("Introduction: the study of divination, present and past," pp. 1–22, and "African divination systems," pp. 194–212) in the book he edited *African Divination Systems: Ways of Knowing*, Indiana University Press, Bloomington, 1991. The article "Hellenophilia versus the history of science," David Pingree, *ISIS*, 83 (1992) 554–563 points to divination as a science. For this and other ideas, the article is highly recommended. Astragalomancy as a forerunner of dice is found in Florence Nightingale David's "Dicing and gaming (a note on the history of probability)," *Biometrika*, 42 (1955) 1–15, as well as pp. 1–27 in her *Games, Gods, and Gambling: The Origins and History of Probability and Statistical Ideas*, C. Griffin, London, 1962, reprinted by Dover Publications, New York, 1998. These include the results of her experiments with astragali which showed that probabilities of the appearance of sides valued at 1, 3, 4, 6 are 1/10, 4/10, 4/10, 1/10, respectively. Hence, the probability of 6, 6, 6, 6 is $(1/10) \cdot (1/10) \cdot (1/10) \cdot (1/10) = 1/10{,}000$. For a discussion of related biblical and Talmudic ideas, see "Random mechanisms in Talmudic literature," A.M. Hasofer, *Biometrika*, 54 (1967) 316–321.

2. A general overview and additional references for divination in the Caroline Islands are in "Divining from knots in the Carolines," William A. Lessa, *J. of the Polynesian Society*, 68 (1959) 188–205. (Also, for a discussion of the spatial models used by the Caroline Island navigators, see Chapter 5 of my book *Ethnomathematics: A Multicultural View of Mathematical Ideas*, Chapman and Hall/CRC, 1994).

In modern mathematics, in $a = b(\text{mod } 4)$, b is defined as the remainder when a is divided by 4. As a result, the values for b are written as 1, 2, 3, 0 where $4(\text{mod } 4) = 0$, rather than, as we have here, $b = 1, 2, 3, 4$.

3. A detailed study of Ifa divination among the Yoruba, including discussion of other sources, is *Ifa Divination: Communication Between Gods and Men in West Africa,* William Bascom, Indiana University Press, Bloomington, 1969. Focusing on Fa, the Dahomey version, is *La Géomancie à l'ancienne Côte des Esclaves*, Bernard Maupoil, Travaux et Mémoires de l'Institut d'Ethnologie, vol. 42. Institute d'Ethnologie, Paris, 1943. Although also discussing Fa rather than Ifa, a valuable discussion of the divination system as one that requires and teaches wisdom and systematic objectivity, is Chapter 11, "Divination and transcendental wisdom," pp. 206–222 in *Ritual Cosmos: The Sanctification of Life in African Religions*, Evan M. Zuesse, Ohio University Press, Athens, 1979.

Sixteen cowries is another form of divination among the Yoruba and their descendants in Brazil and Cuba. In it, the diviners can be both women and men, and the overall process is simpler. Sixteen cowries is related to Ifa in that it is embedded in the same religious system but has a different presiding deity. Each of the outcomes elicits recitation of a set of verses, many of which are similar to the Ifa verses. The point of the divination is a prediction and the associated offering to be made. Here, too, it is the client who selects the verse, and questions phrased as alternatives can be asked to elucidate the verse to be selected. There can, however, only be two alternatives per question as compared to Ifa's 2, 3, 4, or 5. The major difference is that the different outcomes are the number of eyes facing upward when a set of 16 cowries is cast on a tray. There are, therefore, 17 distinct outcomes (0, 1, 2, ..., 16), which are clearly not equally likely. Even if each side of a cowrie shell had the same chance of facing upward, the probability of the outcomes would be $16!/r!(16-r)!2^r$ where $r = 0, 1, \ldots, 16$. Each of the 17 outcomes is identified by a name; 12 of them are similar to the names of the Ifa figures. A thorough discussion of "sixteen cowries" and its relationship to Ifa is *Sixteen Cowries: Yoruba Divination from Africa to the New World*, William Bascom, Indiana University Press, Bloomington, 1980. A fascinating discussion of the training of a "sixteen cowries" diviner is "Schooling, language, and knowledge in literate and nonliterate societies," F. Niyi Akinnaso, pp. 339–385 in *Cultures of Scholarship*, S.C. Humphreys, ed., University of Michigan Press, Ann Arbor, MI, 1997. The article is also recommended because it is more broadly about formal schooling as an educational mode.

5. Several different Malagasy words are used to refer to the diviners and different scholars translate the words somewhat differently. I use *ombiasy* as meaning a diviner who includes healing in his work. Also, to more fully distinguish this mode of divination from other modes, some people specify it as *sikidy alanana*.

For further reading about Madagascar and Malagasy culture, see Nigel Heseltine's *Madagascar*, Praeger Publishers, New York, 1971 and Conrad P. Kottack's *The Past in the Present: History, Ecology, and Cultural Variation in Highland Madagascar*, University Of Michigan Press, Ann Arbor, MI, 1980.

The discussion of sikidy in this chapter is drawn from my more detailed report in "Malagasy Sikidy: A case in ethnomathematics," *Historia Mathematica*, 24 (1997) 376–395. See that article for additional details, specific citations, and additional references. For information about the training of diviners, the formal algorithms, and the checking procedures, I relied, in particular, on *La Divination malgache par le Sikidy*, Raymond Decary, Imprimerie Nationale: Librarie Orientaliste, Paul Geuthner, Paris, 1970; *Pratiques de divination à Madagascar: Technique du Sikily en pays Sakalava-Menabe*, J.F. Rabedimy, Office de la Recherche Scientifique et Technique Outre-Mer, Document No. 51, Paris, 1976; and "Divination among the Sakalava of Madagascar," Robert W. Sussman and Linda K. Sussman, pp. 271–291 in *Extrasensory Ecology: Parapsychology and Anthropology*, Joseph K. Long, ed., The Scarecrow Press, Metuchen, NJ, 1977.

6. The quotation from Boole is on p. 1857 in George Boole, "Mathematical analysis of logic," in *The World of Mathematics*, ed. James R. Newman, 4 vols, Simon & Schuster, New York, 1956, vol. 3, pp. 1856–1858. This is an excerpt reprinted from George Boole's *The Mathematical Analysis of Logic*, Cambridge University Press, Cambridge, 1847.

For a simple discussion of Boolean algebra and switching circuits, see, for example, *Thinking Machines*, Irving Adler, New American Library, 1961 or Chapter 2 in *Mathematical Logic and Probability with Basic Programming*, William S. Dorn, Herbert J. Greenberg, and Sister Mary K. Keller, Prindle, Weber & Schmidt, Boston, 1973. A brief discussion of Shannon's work is on pp. 759–761 in Victor J. Katz, *A History of Mathematics*, HarperCollins, New York, 1993. An example of a text discussing the use of XOR in parity checking is *Computer Engineering: Hardware Design*, M. Morris Mano, Prentice Hall, Englewood Cliffs, NJ, 1988.

7. There is another set of "three inseparables" that are not mentioned in the literature as known to the Malagasy. They are C_2 and C_{16}, C_{11} and C_{13}, and C_{12} and C_{15}.

The directional associations of house placement and use are elaborated in "The Sakalava house (Madagascar)," Gillian Feeley-Harnik, *Anthropos*, 75 (1980) 559–585.

8. For discussions of the historical linkages between Arabic divination, sikidy, and geomancy, see, in particular, "Astrology and writing in Madagascar," by Maurice Bloch, pp. 278–297 in *Literacy in Traditional Societies,*edited by Jack Goody, Cambridge University Press, Cambridge, 1968; *La divination arabe:Études, religieuses, sociologique et folkloriques sur le milieu natif de l'Islam*, Toufic Fahd, E.J. Brill, Leiden, 1966, pp. 196–205; *Les Muselmans à Madagascar et aux Iles Comores*, Gabriel Ferrand, vol. 3, Publications de l'École des Lettres d'Alger, 1902; "L'introduction de la géomancie en Occident et le traducteur Hugo Sancelliensis," pp. 318–353 of vol. 4 of *Mémoires scientifiques*, Paul Tannery, Gauthier-Villars, 1920; and the article preceding it in the same volume, "La géomancie chez les Arabes" by Bernard Carra de Vaux, pp. 299–317. The spread of geomancy throughout Europe is further discussed in *Recherches sur une technique divinatoire: la géomancie dans l'Occident médieval*, Thérèse Charmasson, Centre de Recherches D'Histoire et de Philologie, Librarie Droz, Geneva, 1980.

An analysis of the Arabic version is in Robert Jaulin's *La géomancie: analyse formelle*, Morton & Co., Paris, 1966. This includes mathematical notes by Françoise Dejean and Robert Ferry. I encourage readers of Jaulin's book to also read the critique of it by Marion B. Smith in "The nature of Islamic geomancy with a critique of a structuralist's approach," *Studia Islamica,* 49 (1979) 5–38.

9. A major proponent of the belief that binary choices are fundamental to all human thinking is Claude Lévi-Strauss. See, in particular, his book *The Savage Mind*, University of Chicago Press, 1966, which is translated from the French *La Pensée sauvage*, Librarie Plon, 1962.

CHAPTER 2

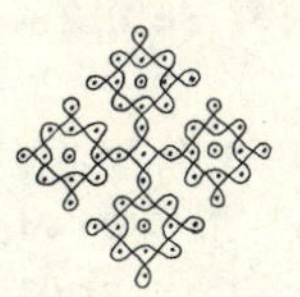

Marking Time

Calendars constructed by different human communities are fascinating in their diversity. What makes them particularly interesting is that they are cultural products often involving religion and/or politics combined with observations of the physical universe. They reflect differing concepts of time and impose different structures on time. Mathematical ideas as fundamental as order, units, and cycles are the very building blocks with which the structures are created. The particular structure that we are taught as children becomes such an intimate part of our life that it is hard to realize that many aspects of the structure are quite arbitrary and that other peoples may perceive, organize, or measure time differently.

The structuring of time can have many functions, some of which are more or less important in different cultures. But everywhere, one of the main functions is to set the schedule of the culture and, thereby, coordinate the activities of individuals in the culture. Other functions may be to relate the group's activities to some natural phenomena or to some supernatural phenomena. The structure may be used to order events in the past or in the future, or to measure the duration of events, or to measure how close or far they are from each other or from the present. Above all, the structure provides a means of orientation and gives form to the occurrence of events in the lives of individuals, as well as in the culture. It provides a continuous and coherent framework in which to mark periodically repeating events and in which to place special events. As such, the structure imposed on time extends well beyond itself, reflecting and affecting much in a culture.

1 We begin our discussion of time structures with some astronomical cycles that have widespread effects on the natural universe. There is a light/dark cycle caused by differing amounts of sunlight received on the earth due to the rotation of the earth on its axis. Human beings can count the number of such cycles; they can arbitrarily divide each cycle into parts; or they can arbitrarily associate groups of cycles together. To count or subdivide the light/dark cycles, there must be an arbitrary point designated as the end of one cycle and the beginning of another.

Another cycle, quite visible to human beings, is the moon going through various shapes in the sky as it revolves around the earth. Again, some arbitrary appearance can be called the beginning of the cycle, and the cycles can be subdivided, counted, or grouped. And, of course, the descriptors of the lunar cycle can be related, in some way, to the descriptors of the light/dark cycle.

The third notable cycle is caused by the earth's revolution around the sun. This cycle is most noticeable from variations in the length of the light/dark cycle, by patterned climatological changes, or by changes in the behavior of flora and fauna. It, too, can be marked as a unit to be counted, subdivided, or related with other like or unlike cycles.

It is significant that we cannot go further in discussing these cycles or specific calendars without establishing some units. That is, we need some structure, no matter how arbitrary, to use as a frame of reference. For this, we introduce and use some basic Western units.

The light/dark cycle was standardized by Western physical scientists into the *mean solar day* (or simply *day*) whose length equals the lengths of the light/dark cycles averaged over a full revolution of the earth around the sun. Its starting point is designated as *midnight*. The day is divided into twenty-four equal subunits called *hours*, which are made up of sixty equal subunits called *minutes*,which, in turn, are made up of sixty equal subunits called *seconds*. A *lunation*, the time from one full moon to the next, does not neatly coincide with light/dark cycles or with mean solar days. On average, in terms of the latter, one lunation equals 29.531 days (29 days, 12 hours, 44 minutes). The *tropical year*, based on the revolution of the earth around the sun, is specified as the time it takes for the apparent sun to return to a particular reference point in the sky. Again, this cycle and its length do not easily match the dark/light cycles or the lunar cycles; its length is 365.2422 days (365 days, 5 hours, 48 minutes, 46 seconds) or 12.368 lunations.

Notice that the *week* has not appeared in the foregoing. That is because it is different in kind: the week has no intrinsic relationship to any physical cycle; it is, instead, a completely arbitrary grouping of some number of days. This difference is of considerable significance. Whereas the calendric concerns of many cultures focus on physical cycles, or often on the reconciliation of different physical cycles, there are others for whom the dominant interest is the interaction of abstract culturally constructed cycles.

In this chapter, we look at some calendars that involve astronomical cycles and their reconciliation. We begin with two different calendars from island groups in the southwestern Pacific--the Trobriand Islanders and the Kodi of Sumba Island. Then, we will look at the Jewish calendar since it is concerned with reconciling the same astronomical cycles but does so in a very different manner. The Jewish calendar, in addition, includes an arbitrary 7-day cycle. In the next chapter, we turn our attention to calendars dominated by arbitrary, abstract cycles.

2 The Trobriand Islands lie off the eastern coast of Papua New Guinea (see Map 2.1). Over time, interaction with others has modified the culture of the Trobriand Islanders. Here, however, using the present tense, we discuss the culture of the Trobriand Islanders as it existed in the first half of the twentieth century.

About 50% of the Islanders' time is devoted to agricultural pursuits. The planting and care of their gardens are a major focus of their lives. Gardening is surrounded by many rites, and the neatness and aesthetics of a garden are of as much concern as its yield. The garden cycle consists of planting, fencing, weeding, harvesting, cutting, and burning the fields to ready them for the next planting, and a period when no garden work is done. The organization of gardening, that is, designating the beginning of each activity, is a major concomitant of political power.

To the Trobriand Islanders, the moon is of particular importance. Although their concern is with seasons, their calendar is based primarily on lunar cycles. Without the light of the moon during the dark of night, activities are confined to the home, but with moonlight, outdoor activities are possible. Hence, the part of the lunar cycle around the full moon is their major interest. The lunar cycle begins with the first appearance of the moon. This phase, until the moon is overhead at sunset, is "the unripe moon." This is followed by "the high moon." Starting on the tenth day, the days are specifically named. The days of the full moon, what we would call the thirteenth, fourteenth, and fifteenth, are the times for evening festivities with socializing, dancing,

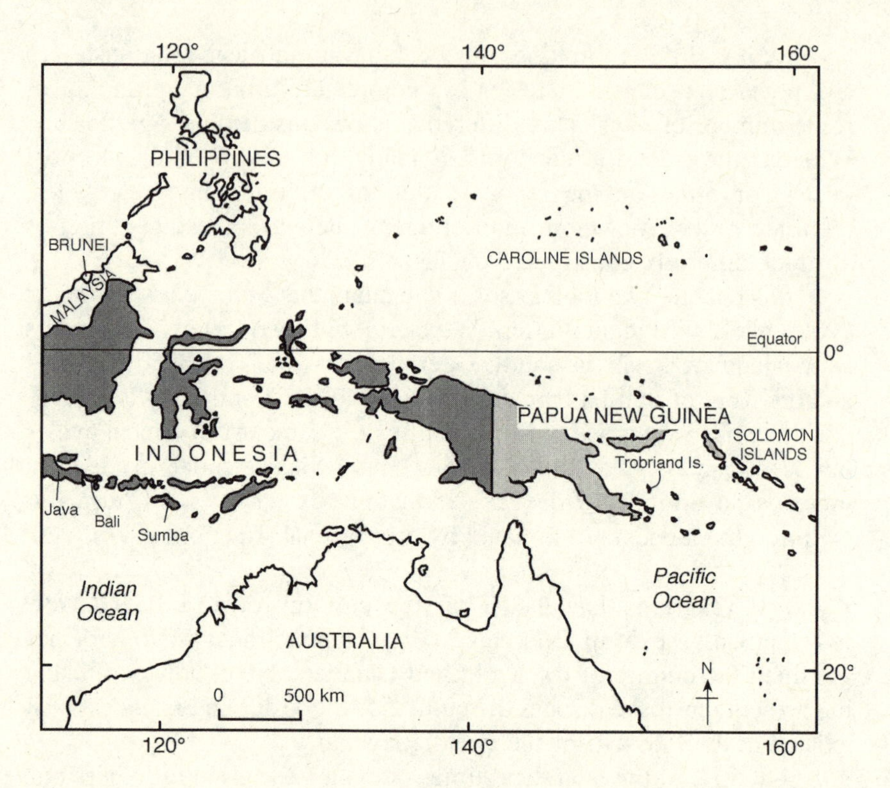

Map 2.1 Papua New Guinea and Indonesia

and gift-giving. Starting with the twenty-second day, it is "in the great darkness," and the days are again unnamed.

The Trobriand structuring of time is intimately related to gardening. A year is a full garden cycle. The past year is literally "the time of the past yam." Also, to place an event some years ago, one recalls which set of fields was planted and reconstructs the time passage based on a cyclic pattern of field planting. There are 29 or 30 days in each of the lunar cycles (we will call them months), and there are 12 or 13 months in a Trobriand year. Most of the months are named, but which named month it is differs in different locations within the Trobriands. What is special about the calendar is how the names and number of months vary from place to place and within a year.

The Trobriand yearly cycle has no specific point designated as its beginning or end. However, we will call their month of Kuluwasasa the last month because it is the month during which the harvest takes place, and the elders meet to formally decide on plans for the community's activities during the next garden cycle. In terms of their staple crops, the Trobriands form four different districts, which have different harvest times. The month of Kuluwasasa, however, is *always* the harvest time, and so it occurs first on the outlying island of Kitava, next on the southern end of the main island of Kiriwina, then on the northern end of Kiriwina, and finally on the island of Vakuta. The next month name is Milamala. Since everyone knows that the months are regionally offset and the way they are offset, people on the southern end of Kiriwina might, for example, call their Kawal (which follows Milamala) "Kitava's Milamala" instead. Milamala is emphasized because it is the month when the spirits of the departed visit their villages, and, at the time of the full moon, when the spirits are expelled, there is a major festival with much feasting and dancing. This festival also marks the beginning of the burning phase of the garden cycle.

In order for a calendar based on the lunar cycle to stay in synchronization with the seasons that are sun-related phenomena, there must be a way to vary the number of lunar cycles per year so that most years have twelve lunations, but every third year or so, there are thirteen. (Recall that a tropical year equals 12.368 lunations.) The Trobriand Islanders do that by using a method that demonstrates that valid calendars need not involve precise mathematical calculations, extensive record keeping, or elaborate astronomical knowledge. Their method is to rely on the internal clock of a biological organism. A particular marine annelid spawns just once each tropical year, at the time of a full moon, in the sea off the island of Vakuta. If the worm, called *milamala* by the Trobriand Islanders, does not appear at the full moon of Vakuta's month of Milamala, the festival is delayed, and the month is repeated–that is, there is a second month of Milamala that causes that year to have 13 months. Since the other regions have already had their Milamala festivals and are already in a later phase of the garden cycle, for them, the doubling up of a month occurs later but before the next Milamala. Exactly when they add a month is not clear, perhaps, in part, because referring to the months toward the end of the year by name is not of major concern.

With or without a calendar, the round of seasons is evidenced by periods of calm or heavy winds, rain, and dryness. The gardening activities, so important to the Trobriand Islanders, *must* correlate with these natural phenomena. But these alone do not determine when events take place because the gardening cycle must also correlate with religious rites and festivals. When, for example, Vakuta's Milamala festival is deferred for a month, the people cooperate to speed up the cleaning and burning part of the garden cycle so that it is sure to be done by the end of the dry season. And even more significant, although the regional gardening activities are asynchronous, the calendric structuring of activities is maintained by having a calendar essentially made up of a supracycle consisting of four almost identical subcycles offset from each other by one month. Clearly, this calendar is not intended to measure elapsed time or to assign a precise label to each past or future day. It does, however, order and coordinate the activities of those within the group as well as link the activities of the groups to natural and supranatural phenomena.

At about the same latitude (10°), but some 3200 km away, on the island of Sumba in Indonesia, quite similar calendars are in use. They are similar in focusing on lunar cycles with adjustment to remain synchronized with the seasons. Since the cultures are different, the events that make up the cycles of activities are also different. According to the myths of the Kodi (a Sumba Island group of about 50,000 people), time units became noted and named only after their ancestors arrived on the island. When their ancestors arrived, people lived a continuous cycle; they were young, grew old, and then were young again. The sun was close to the earth, and the moon was steady in the sky. But a house was built to be taller than the rest, and the builder asked the Creator to raise the sun to avoid burning his roof. The sun was raised, but the price was periodic light and darkness. Two brothers, trying to see whose spear could be thrown higher, hit the moon and broke it. They threw the broken pieces of moon back into the sky at sunset, and after that, the moon followed a cycle of waxing and waning. With these divisions of time came mortality. Then time units became noted and named because now time counted.

There is, then, the day/night cycle, but when counts between events are made, the counts are of the number of intervening nights. The moon is all important because it lights the darkness of the night. For the Kodi, as for the Trobriand Islanders, with the full moon comes the time for feasts, night dancing, and socializing. The lunar cycles are named and noted, as are the two marked seasons: dry and rainy. Here, too, within the gardening cycle are times of harvesting, burning the fields, planting and weeding, and waiting for the crops to mature. The ritual cycle includes "the bitter period" lasting about four months, during which activities that might endanger the growth of the plants are prohibited:

no children's games, no making noise or music, no tattooing, and no spearing pigs and buffalo. When these prohibitions are lifted, there is a return to gaiety and romance and preparations for a major festival. During the festival, thousands of riders in ceremonial clothes participate in fierce jousting competitions. But upon asking how to find out about the structure holding all of these together, a Western visitor was told by a Kodi elder to "...start with the sea worms. That is where we start ourselves."

The highest ranking Kodi priest, the *Rato Nale* (Priest of the Sea Worms) is responsible for the calendar. It is he who announces when the bitter period ends and, a few weeks later, that people should start preparations for the festivities because the worms will appear in seven nights at the full moon. (These are the same marine annelids of the Eunicid family, *Leodice viridis*, that appear on the Trobriand Islands.) His announcement is based on his knowledge of the moon and seasonal indicators in the environment. Here, as contrasted to the Trobriands, the worms' appearance is predicted rather than used as an after-the-fact correction of the calendar. The intercalation of a month is probably made by the Rato Nale toward the end of the dry season and before the bitter period. At that time, the names of the months are not of much interest; in fact, there is mention of a month that "has no name" and a period of "forgetting the moon name." With the Rato Nale's announcement of a festival ushering in the time of bitter sacrifices, month names become specified and of interest. The role of the Rato Nale, however, is more than "to count the days and measure the years." He serves as an exemplar of ritual discipline, and as such, he interacts with, and controls, the natural and cultural phenomena that are synchronized through the calendar. During the rainy season, after the crops are planted, for example, he sits for more than a month relatively immobile and concentrating. He cannot leave his house or enter the women's part of the house and special food must be prepared for him in a special area. Were he to leave the house, extreme winds would blow and harm the young plants, and were he to eat taboo foods, lightning would strike the fields. Tidal waves would result if he did not coordinate certain events. His behavior, in short, is part of what maintains the balance, coherence, and unity that the calendar is intended to provide.

Severe problems have arisen since Indonesian independence from colonial powers because Indonesian government officials want to know *when* the festivals will occur. They insist on fixed dates in the government calendar, specified well in advance. This depends on formalized and regularized measures of elapsed time, which the Kodi calendar does not include. Even if elapsed time between events were a feature of the Kodi calendar, the translation from one time structure to another would still be a far more complex problem than is commonly realized.

The calendars of the Trobriand Islanders and of the Kodi are luni-solar calendars; that is, they focus on lunar cycles but, with the aid of the worms, remain in phase with the seasons and with the solar cycles. Their reliance on the internal clock of the worm brings forcefully to our attention the interplay of biological and astronomical phenomena. But, even more so, it demonstrates a non-technical, but effective, means of creating a luni-solar calendar.

3 We turn now to the Jewish calendar, another luni-solar calendar, in which quite different techniques are used to reconcile the lunar and solar cycles. In this case, the calendar is quite formal in structure; explicit calculation rules are used to determine when each year and month begins, and each day carries a specific label as to its place in the month and year. Further, as an integral part of it, the Jewish calendar has a 7-day cycle (week), and there is a designated starting point that enables the years to be numbered and elapsed time to be measured.

Unlike the Trobriand Islanders and Kodi, the Jews are geographically and culturally diverse. While, historically (and perhaps even currently for some of its communities), the calendar was related to agricultural practices, through time, it has primarily become one of the major unifiers of a broadly dispersed people. The calendar coordinates their religious observances. It is intertwined with the Bible, which sets forth the observances and, for many of the observances, specifies *when*, as well as how, they should be carried out. The Jewish people are described as *the people of the Book*; the calendar both operationalizes and reinforces many aspects of *the Book*. For example, the annual rereading of the Book—when it begins, which portions to read, and when it ends—are all included in the schedule, which the calendar structures. The stipulations that there be counting of years and labeling of days and months, and that the calendar incorporate lunar cycles, solar cycles, and the 7-day week, are found in the Bible. The details of the calendar structure, however, were left to be worked out by practitioners of the religion.

Up to about 1600 years ago, a designated group of Jewish elders, called the Sanhedrin, annually set the specific configuration of each year. Their decision was sent by messengers to the widely dispersed Jewish population. When the persecution of the Roman Emperor Constantine interfered with these messengers, the Jewish scholar Hillel II made the calculation rules public. Even though public, the rules were not widely understood. Although there have been minor changes, the calendar has remained essentially the same to this day.

The ritual day starts at sunset and ends with the appearance of three stars on the following evening. However, for the purposes of calendric calculations, a day consists of 24 equal hours, each of which consists of 1080 halakim, and the new day is said to begin at what we now call 6 p.m. A lunar cycle, as was previously noted, is just over 29.5 days. In the units of this calendar, it is considered to be 29 days, 12 hours, and 793 halakim. Thus, months are set at either 29 or 30 days.

As with other luni-solar calendars, in order to stay in synchronization with solar, as well as lunar, cycles, some years must have 12 months and some 13 months. In the Jewish calendar, this leap year/non-leap year pattern forms a 19-year cycle. The significance of 19 years is that 19 solar years $\approx$ 235 lunations, or, more exactly, 235 lunations exceed 19 solar years by about 4.5 hours. Thus, 19 solar years are quite close to 19 lunar years when 12 of the lunar years have 12 months each, and seven of them have 13 months each ($12 \cdot 12 + 7 \cdot 13 = 235$). Within the 19-year cycle, the years that are leap years (that is, those with 13 months rather than 12) are the third, sixth, eighth, eleventh, fourteenth, seventeenth, and nineteenth (see Figure 2.1). The starting point of the calendar is at the beginning of a 19-year cycle, and so the place of a year in the cycle is found by calculating the year number (mod 19), that is, the remainder when the year number is divided by 19.

There are, however, just 12 month names (see Table 2.1). Although the month of Tishri begins the year (around the autumnal equinox) and Elul ends it, originally, Nisan (around the vernal equinox) was referred to as the first month and Adar as the last month. Thus, as with the Trobriand Islanders, the year was extended by repeating the name of a month; the repeated month was the last in the yearly cycle. (When two months of Adar are present, they are now sometimes distinguished by being called Adar I and Adar II. Clearly, it is Adar I that is the inserted month as the religious observances that take place annually in Adar are in Adar II. An example of an observance in Adar is the festival of Purim which is on the 14th of Adar. The number of festivals that fall at mid-month, the time of the full moon, is also noteworthy.)

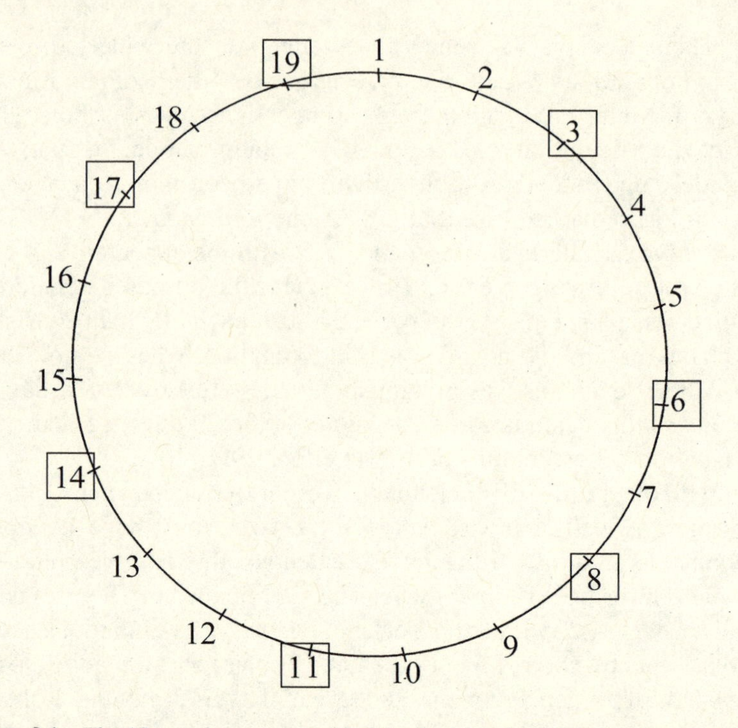

Figure 2.1 The 19-year leap year cycle of the Jewish calendar. Leap years are shown in boxed numbers. (Notice that the spacing in the cycle is symmetric with respect to year 3. Moving in either direction, the spacing is 3, 2, 3, 3, 3, 2, 3.)

Table 2.1 Months in the Jewish calendar and the number of days in each

Tishri	30
Heshvan	29 (or 30)
Kislev	30 (or 29)
Tevet	29
Shevat	30
[Adar	30 (this month is present only in leap years)]
Adar	29
Nisan	30
Iyyar	29
Sivan	30
Tammuz	29
Av	30
Elul	29

Of fundamental importance to the Jewish calendar is the 7-day cycle. The recognition of the Sabbath, the seventh day of the cycle, as a day of no work devoted solely to religious and contemplative activities, is a major tenet of the religion. It is of such importance that it appears as part of the story of creation and is one of the ten commandments. The idea of a 7-day unit probably was adapted from the Babylonians and Assyrians during the period about 2500 years ago when the Jews lived among them. But the concept of the Sabbath and its centrality were added, and the Sabbath became crucial to the structure of the calendar.

The feature of the calendar that makes it appear quite complex and which involves special rules is that the *dates* (month name and day number within the month) of holidays are fixed, but the lengths of some months and, hence, the lengths of the years are not. Here, the Sabbath plays a crucial role. It constrains the *days* on which certain religious dates may fall thereby becoming involved in the determination of the start of the year and the number of days--29 or 30--in the months of Heshvan and Kislev. Tishri 10, for example, cannot fall on a Friday or Sunday, the days before or after the Sabbath. The day, called *Yom Kippur*, is one of fasting and very important religious observances. If it were to fall on a Friday, no food could be prepared for the Sabbath and so it, too, would, de facto, become a day of fasting. Similarly, if it fell on a Sunday, the fast would be extended through Monday because no preparations were possible for two consecutive days. In addition, Tishri 21 (*Hoshanah Rabbah*) cannot fall on the Sabbath; part of its special observance requires carrying around some objects, which contradicts a Sabbath prohibition. Combining these restrictions on Tishri 10 and Tishri 21 means that the first day of the year, Tishri 1, cannot be a Wednesday, Friday, or Sunday. This, of course, also means that a year cannot end on a Tuesday, Thursday, or Sabbath. Thus, lunar cycles, regular or leap years, and what the *day* is are intermingled in the setting of Tishri 1. Then, the same considerations are used to set Tishri 1 for the following year. Once *both* are known, the month lengths are adjusted to make one year end where the other begins.

According to Jewish belief, creation and the new moon of Tishri of Year 1 were at 5 hours and 204 halakim after the start of a Monday. (That corresponds to 11 minutes and 20 seconds after 11 p.m. on a Sunday evening.) Hence, to find the day and time of the new moon some years later, one adds to this start time the product of the number of lunar cycles that have passed and the length of a lunar cycle. For example, keeping leap years in mind, the first new moon of the year

5755 is 302 full 19-year cycles plus 5 leap years plus 11 ordinary years after the creation; that is:

$$302\,(235) + 5\,(13) + 11\,(12) = 71,167 \text{ lunar months later.}$$

$$\begin{aligned} &71,167 \text{ lunar months} \\ &\quad = 2,099,426 \text{ days} + 12 \text{ hours} + 56,435,431 \text{ halakim;} \\ &\quad = 2,101,603 \text{ days} + 19 \text{ hours} + 31 \text{ halakim;} \\ &\quad = 300,229 \text{ weeks} + 19 \text{ hours} + 31 \text{ halakim.} \end{aligned}$$

Since only the day and time are being sought, full weeks can be disregarded. Adding 19 hours and 31 halakim to the *day* and *time* of creation, the result is 24 hours and 235 halakim after the start of a Monday or 235 halakim after the start of a Tuesday. (That corresponds to just after 6 p.m. on a Monday evening.)

The rules for setting the day of Tishri 1 for any year are:

Tishri 1 is the day of the new moon of Tishri except if:

(1) the day is a Wednesday, Friday, or Sunday; or
(2) the time of the new moon is 12 noon or later; or
(3) it is not a leap year and the new moon is a Tuesday at 204 halakim after 3 a.m. or later; or
(4) it is a year following a leap year and the new moon is on a Monday at 589 halakim after 9 a.m. or later.
In these cases, Tishri 1 is on the following day unless that day is a Wednesday, Friday, or Sunday, in which case, it is the day after that. These rules are shown diagrammatically in Figure 2.2.

In our example of the year 5755, the new moon is just after the start of Tuesday, which is affected by none of the above exceptions, and so Tishri 1 is a Tuesday. To find the new moon of Tishri for 5756, since 5755 is a leap year, another 13 lunations [that is, 13 (29 days+12 hours+793 halakim)] are added to the new moon time of 5755:

$$\begin{aligned} &13\,(29 \text{ days} + 12 \text{ hours} + 793 \text{ halakim}) \\ &\qquad +235 \text{ halakim after the start of Tuesday} \\ &\quad = 54 \text{ weeks} + 5 \text{ days} + 21 \text{ hours} \\ &\qquad +824 \text{ halakim after the start of Tuesday} \\ &\quad = 21 \text{ hours} + 824 \text{ halakim after the start of Sunday} \\ &\quad = 824 \text{ halakim after 3 p.m. on Sunday.} \end{aligned}$$

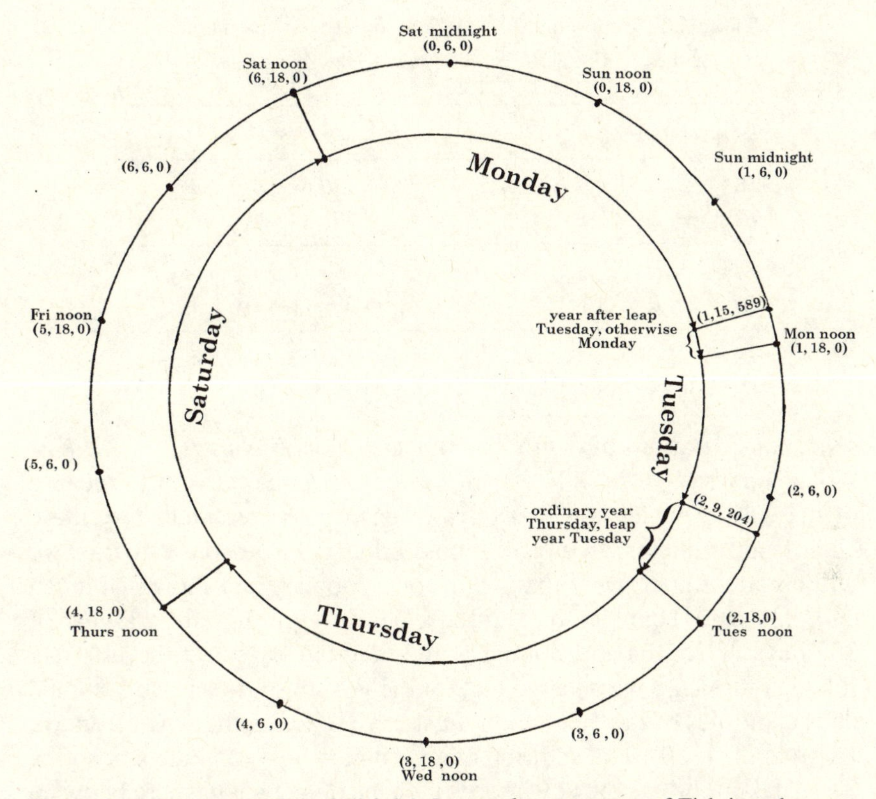

Figure 2.2 Finding the day of Tishri 1. Locate the new moon of Tishri on the outer circle. Times are expressed in terms of days and hours from Saturday at 6 p.m. Where there are parts of hours, they are measured in halakim. [For example: (0, 0, 0) is 6 p.m. on Saturday; (0, 18, 0) is noon on Sunday; (4, 0, 0) is 6 p.m. on Wednesday; and (2, 9, 204) is 204 halakim—about 11 minutes and 20 seconds—after 3 a.m. on Tuesday.] The inner circle shows the corresponding day for Tishri 1.

According to both the first and second conditions then, Tishri 1 of the year 5756 is on a Monday.

Now that we know that the year 5755 is a leap year beginning on a Tuesday and ending on a Sunday, we are finally ready to select the length of the year and the number of days in Heshvan and Kislev. A normal year is considered to be one in which Heshvan is 29 days and Kislev is 30 days. To decrease the number of days in a year, Kislev also becomes 29 days, and to increase it, Heshvan becomes 30 days. Hence, a leap year can have 383 days ($H = 29$, $K = 29$), 384 days ($H = 29$, $K = 30$), or 385 days ($H = 30$, $K = 30$) (see Table 2.2). To begin a year on a Tuesday and end it on a Sunday, the choice must be the 384-day year with $H = 29$, $K = 30$. If this were not a leap year, there would be the

Table 2.2 Year lengths in the Jewish calendar [the number of days in the months of Heshvan (H) and Kislev (K) are also shown]

Leap years	Regular years
383 = 54 weeks + 5 days ($H = 29$, $K = 29$)	353 = 50 weeks + 3 days ($H = 29$, $K = 29$)
384 = 54 weeks + 6 days ($H = 29$, $K = 30$)	354 = 50 weeks + 4 days ($H = 29$, $K = 30$)
385 = 55 weeks ($H = 30$, $K = 30$)	355 = 50 weeks + 5 days ($H = 30$, $K = 30$)

same considerations but with the choices being 353 days ($H = 29$, $K = 29$), 354 days ($H = 29$, $K = 30$), or 355 days ($H = 30$, $K = 30$). Because of the way in which the first days of each year are calculated, these choices are sufficient to cover all possibilities. To demonstrate this, we will use an ordinary year that begins on a Monday and show that it can end only on a Thursday or Saturday, hence requiring either a year of 353 days or 355 days. Looking at the diagram based on the first day rules (Figure 2.2), we observe that for the year to start on a Monday, the days and times of the new moon must have fallen in the range from (6, 18, 0) to (1, 18, 0). The addition of an ordinary year (354 days, 8 hours, 876 halakim) brings the new moon of the next year somewhere between (4, 2, 876) and (6, 2, 876), which, again looking at the diagram, means that it would be a Thursday or a Saturday.

Once the number of months in a year (12 or 13), the start day (Monday, Tuesday, Thursday, or Saturday), and the number of days in each of Heshvan and Kislev (29 or 30) have been determined, everything necessary is known. All the days of religious observance and festivities have their fixed calendar dates; that is, Tishri 1 is the New Year, Tishri 10 is Yom Kippur, Tishri 21 is Hoshanah Rabbah, and so on. They will fall when they should with respect to the Sabbath, the moon, and the season. (A summary of the algorithm for determining the specifics of year Y is in Appendix I of this chapter.)

4 With the foregoing examples in hand, let us pause and reflect on some calendric issues. Calendars, in general, have operational and/or intellectual levels; that is, for example, some calendars focus on organizing ritual and agricultural activities, while others are also, or instead, concerned with structuring the flow of historical events.

Another categorization is that calendar structures may incorporate components that are environmental, and/or components that are related to social structure, and/or components that have psychological concomitants.

A major issue is whether people believe time to be cyclic or linear. However, circularity and linearity are not necessarily mutually exclusive; nor, if they exist together, does one have to be subordinate to the other. Just as an automobile tire can rotate as the center of the hub moves linearly forward with respect to a road, cycles of time can move along lines. Or, points along a piece of thread can be thought of as in linear order, but can be viewed simultaneously as cyclically related if the thread is wrapped around a central spool. Also, where the belief is in several cycles, the cycles can be independent, one within another, or linked in some other way.

Some cultures integrate all calendric functions into one system, while others keep them distinct. Overall, however, it is not only what components or configurations are included, but how they are mixed and matched. And since our primary interest is mathematical ideas, we particularly note that calendars differ in their formality, public accessibility, and whether or not they can be described with clear, algebra-like algorithms.

In just the few calendars we have already examined, many of the foregoing components are apparent. In all of the calendars, there have been environmental and social components, including agricultural and ritual structuring. The Jewish calendar differs from the other two by incorporating a means of pinpointing historical events and measuring elapsed time. Cycles dominate all of them, but the Jewish elapsed time measure incorporates a linear aspect as well. The measure also implies that time is continuous (there are no gaps in it) and uniform (the size of the units does not change because they are at some time rather than another).

We have, thus far, focused only on luni-solar calendars. There are, of course, many other types. The Muslim Hijra calendar, for example, is solely lunar; that is, it marks the lunar cycles and remains in synchronization with them. The Gregorian calendar (the official calendar of the U.S.A. and of many other nation-states), however, is strictly solar, marking and remaining in synchronization only with solar cycles. (Although the Gregorian calendar has within it units referred to as *months*, these are historical vestiges and do not coincide with lunar cycles.) Both the Hijra and Gregorian calendars have a 7-day cycle (*week*), but in each, the cycle is independent of the other aspects of the calendar structure rather than integrated with them. The Jewish, Hijra, and Gregorian calendars are all formal, public, and algorithmic, and each enables measuring elapsed time by virtue of having a specific starting point, imposing units on the flow of time, and, thus, explicitly and uniquely identifying each day and each instant within it. Because they share these features, the Hijra, Jewish, and Gregorian calendars can be translated, one from the other, but since the lengths of their months and years differ, and their starting points differ, that translation involves understanding, and working with, all their different structures.

Today, many people live under more than one calendar--calendars particular to their religious or cultural groups and calendars associated with their nation-states. The domination of one group by another has frequently been marked by the imposition of the dominator's calendar. In some cases, the indigenous calendar was completely replaced, in others, it was modified, and in many cases, the two continued to exist side by side. Even within the same culture, new rulers often decreed new or modified calendars to mark their coming to power or their adoption of a new ideology. And the spread of commerce, travel, and communication led many people to adopt calendars more widespread than their own, to be used especially for purposes of external interactions.

Expressing one calendar in terms of another is a very difficult undertaking, which, depending on the calendars, may not even be possible. The calendars differ not only in detail but in underlying structures as well. They differ in intent as well as in the activities they are meant to coordinate. How to live with different calendars begins with understanding the structure and purposes of each and with understanding that none is more advanced or more correct. Within one's own life, the actual challenge of living under multiple calendars is the challenge of reconciling the different cultural priorities expressed in each.

Notes

1. The *second*, as described here, is 1/86,400 of the mean solar day. In 1967, the General Conference on Weights and Measures, in order to establish a standard independent of the precision of astronomical measurements, redefined the second to be 19,192,631,770 periods of the radiation in the hyperfine transition of the cesium-133 atom. This change represents a marked conceptual shift to a system that is no longer astronomical. It is an excellent example of the way that "objective" standardization moves further and further from the daily experiential world. The minute is now 60 of these atomic seconds, the hour 60 such minutes, and the day 24 of these hours. For specifics of the detailed issues involved, see the section "Time as systematized in modern scientific society," pp. 668–673 in *The New Encyclopedia Britannica*, vol. 28, Encyclopedia Britannica, Inc., Chicago, 15th edition, 1993.

2. My discussion of the Trobriand Islanders and their calendar is based on Bronislaw Malinowski's "Lunar and seasonal calendar in the Trobriands," *Journal of the Royal Anthropological Institute of Great Britain and Ireland*, 57 (1927) 203–215, and his *Coral Gardens and Their Magic*, vol. 1, Indiana University Press, Bloomington, IN, 1965 (original edition 1953); and on "The seasonal gardening calendar of Kiriwina, Trobriand Islands," Leo Austen, *Oceania*, 9 (1939) 237–253; and "Primitive calendars," Edmund R. Leach, *Oceania*, 20 (1950) 245–262.

The book from which the Kodi information is adapted, *Play of Time: Kodi Perspectives on Calendars, History, and Exchange*, Janet Hoskins, University of California Press, Berkeley, CA, 1993, is particularly recommended. The phrases quoted are from pages 80, 335, and 336.

3. The calendric stipulations in the Bible can be found, in particular, in Numbers 28 and 29, Deuteronomy 16 and Leviticus 23. In these, the observance that is now considered to start the new year is described as being on the first day of the *seventh* month. Here, that is considered to be the *first* month. Since the months cycle, none of the relative positions of other observances are modified. The only modification that has been made to the biblical stipulations is the starting point of the continuous cycle of months. As it currently stands, the starting point of the annual cycle coincides with the time of creation, and it begins the annual rereading of the Bible and a new year number.

Here, for the convenience of the reader, we use the day names Sunday, Monday, Tuesday, Wednesday, Thursday, Friday, Saturday (Sabbath), rather than the designations first day, second day..., seventh day (Sabbath), as in the Jewish calendar. References to discussions of day names are in Note 1 of the next chapter.

For useful specific discussions of the Jewish calendar, see *Jewish Calendar Mystery Dispelled* by George Zinberg, Vantage Press, New York, 1963 and V.V. Tsybulsky's *Calendars of the Middle East Countries*, U.S.S.R. Academy of Sciences, Institute of Oriental Studies, Nauka Publishing House, Moscow, 1979. The latter includes several other calendars such as the Muslim Hijra lunar calendar, the Coptic–Egyptian calendar, the Iranian solar Hijra calendar, and several calendars from Turkey. This book also has extensive tables for correlating calendars. An excellent book, including the Jewish calendar and thirteen others is *Calendrical Calculations*, Nachum Dershowitz and Edward M. Reingold, Cambridge University Press, New York, 1997. The stated goal of the authors is unified, algorithmic presentations suitable for LISP implementation. (The implementations are included.) Their presentation is, of course, limited to calendars that can be described in this way. For a fascinating historical note, see the article "Al-Khwārizmi on the Jewish calendar," E.S. Kennedy, *Scripta Mathematica*, 27 (1964) 55–59 which sketches the contents of a ninth-century treatise by an Arabic scholar. The treatise describes the rules of the Jewish calendar and calculations used. For a more general discussion of the Jewish calendar and its history, see "Calendar, History of," pp. 498–501, and "Calendar," pp. 501–507 in *The Jewish Encyclopedia*, vol. 3, Isidore Singer, ed., KTAV Publishing House, New York, 1964. An extensive bibliography can be found in "Gregorian dates for the Jewish New Year," Edward L. Cohen, pp. 79–90 in *Proceedings of the Canadian Society for the History and Philosophy of Mathematics—22nd Annual Meeting*, vol. 9, J.J. Tattersall, ed., 1996.

4. For discussions of the Muslim Hijra lunar calendar, see *Calendars of the Middle East Countries* and *Calendrical Calculations*, both cited in Note 3 above. For the Gregorian calendar, see those books and also "Mathematics of the Gregorian calendar," V. Frederick Rickey, *The Mathematical Intelligencer*, 7 (1985) 53–56. Using the dissociation of Easter from Passover as a case study, in "Easter and Passover: on calendars and group identity," *American Sociological Review*, 47 (1982) 284–289, Eviatar Zerubavel examines the way groups use calendars to distinguish themselves from other groups.

Some interesting theoretical discussions of calendars and time are "Lineal and nonlineal codifications of reality," Dorothy Lee, *Psychosomatic Medicine*, 12 (1950) 89–97; "Two essays concerning the symbolic representation of time," pp. 124–136 in *Rethinking Anthropology,* Edmund R. Leach, Athlone Press, London, 1961; "Primitive time-reckoning as a symbolic system," Daniel N. Maltz, *Cornell Journal of Social Relations*, 3 (1968) 85–112; and "Theories of time," Chapter 7, pp. 147–156 in *Time, Experience and Behaviour*, J.E. Orme, Iliffe Books, London, 1969. Combining the theoretical with the specific are "Time and historical consciousness: The case of Ilparakuyo Maasai," Peter Rigby, pp. 201–241 in *Time: Histories and Ethnologies*, Diane O. Hughes and Thomas R. Trautmann, eds., University of Michigan Press, Ann Arbor, MI, 1995, and "Indian time, European time," Thomas R. Trautmann, pp. 167–197 in the same book. *Cultures and Time*, Louis L. Gardet et al., UNESCO Press, Paris, 1976 is a collection of articles on the empirical perception of time and the conceptions of time and of history in India, among people of the Maghreb, in Jewish culture, in Chinese thought, and in Bantu thought.

Appendix I. Summary of algorithm for year Y

Constants:

C = time of creation = 5 hours + 204 halakim after start of Monday
L = length of 1 lunation = 29 days + 12 hours + 793 halakim
W = length of 1 week = 7 days = 168 hours = 181,440 halakim

Note: For calculations, a day starts at 6 p.m. on previous evening.

I. Determine if year Y is a leap year or a regular year.

Use $r = Y \pmod{19}$ and Figure 2.1. Note that $Y = 19Q + r$ where Q is the number of full 19-year cycles since creation.

II. Find D, the day of Tishri 1 in year Y.

(a) Calculate N = number of lunations since creation:

$$N = 235Q + 13a + 12b$$

where a and b are the number of leap years and regular years respectively since the start of the current 19-year cycle. (Note that $a + b = r - 1$.)

(b) Calculate T, the day and time of the new moon of Tishri:

$$T = C + LN \pmod{W}.$$

(c) Find D by applying rules in Figure 2.2 to T.

III. Find D' the day of Tishri 1 in year $Y + 1$.

(a) Calculate T', the day and time of the new moon of Tishri:
If year Y is a leap year,

$$T' = T + 13L \pmod{W};$$

If year Y is a regular year,

$$T' = T + 12L \pmod{W}.$$

(b) Find D' by applying rules in Figure 2.2 to T'.

IV. Find the length of year Y and the number of days in Heshvan and Kislev.

Use D'–D, whether year Y is a regular or leap year, and Table 2.2.

Example: $Y = 5755$

I. $5755 = 19 \cdot 302 + 17$; $r = 17$, $Q = 302$, year is a leap year.

II. (a) $N = 235(302) + 13(5) + 12(11) = 71{,}167$ lunar months.

(b) $T =$ (5 hours + 204 halakim after start of Monday) + 71,167 (29 days + 12 hours + 793 hal) [mod weeks] = (5 hours + 204 halakim after start of Monday) + 300,229 weeks + 19 hours + 31 hal [mod weeks] = 24 hours + 235 halakim after start of Monday = 235 hal after start of Tuesday.

(c) $D =$ Tuesday

III. (a) $T' = T + 13L \pmod{W}$ = 235 hal after start of Tuesday + 13 (29 days + 12 hours + 793 hal)[mod weeks] = 54 weeks + 5 days + 21 hours + 824 hal after start of Tuesday [mod weeks] = 21 hours + 824 hal after the start of Sunday = 824 hal after 3 p.m. on Sunday.

(b) $D' =$ Monday

IV. D'–D = Monday–Tuesday = 6 days; year is leap year so length of year = 384 days and Heshvan = 29 days, Kislev = 30 days.

Result

Year 5755 begins at sunset on a Monday evening and has 13 months with a total of 384 days; Heshvan and Kislev are 29 and 30 days, respectively.

CHAPTER 3

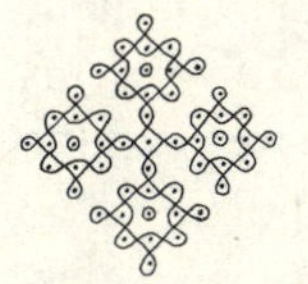

Cycles of Time

In this chapter, we continue with calendars, but highlight a different aspect, one that has been, and continues to be, of particular importance to many peoples in the world. For many people, the crucial role of a calendar is determining the *quality* of time. Knowing where a day is situated in one or more cycles carries with it knowledge about how the day is related to the culture's cosmology, how to interpret the day, or what to expect of the day.

1 Most people are familiar with the 7-day cycle, commonly referred to as a *week*. It is a culturally established cycle, not determined by some happening in the physical world. The broad spread of the 7-day cycle is frequently attributed to its Babylonian origin, where it was associated with the zodiac. Then, picked up and adapted by people from India, it spread with their influence and the influence of the Hindu religion. The Jewish people, upon leaving Babylonia reinterpreted the 7-day cycle into a central tenet of their religion. As such, it was passed on by them to the Christians and to the Muslims. Together, these groups influenced a goodly number of cultures throughout the world. Despite its ubiquity, a 7-day cycle is no more rational than, say, a 6-day cycle or an 11-day cycle. And, just as a single culturally determined cycle can be part of a calendar, so several such cycles can be parts of a calendar. Within a particular culture, some arbitrary cycle may, or may not, have developed from market cycles, or visiting patterns, or numerous other human or non-human affairs. None the less, once the cycles become part of a calendric structure, they become abstract cycles, often interacting with other calendric cycles and, yet, still resonating with other aspects of the culture.

The mathematical ideas surrounding culturally determined cycles differ from the ideas involved with physically derived cycles. The central issue in the previous chapter was that the calendar remains in synchronization with some aspect of the physical world. To this end, leap months and/or leap days were added. The history of the Gregorian calendar, for example, revolves around adjusting the occurrence of leap days to better and better fit the solar cycles. The problem stemmed from the fact that the length of the solar cycle is not expressible as a whole number of days; the increasing accuracy of scientific observation was linked to modifications in the calendric computations. This issue does not arise with arbitrary, abstract cycles that are created at will. For them, instead, the issue is usually how to interrelate cycles made up of differing numbers of days. The difference is what, in modern mathematical terms, we might categorize as problems in pure mathematics versus problems in applied mathematics.

Let us look first at correlating full-day cycles of different lengths. For example, a cycle of 3 days (D1, D2, D3) and an independent cycle of 5 days (d1, d2, d3, d4, d5) will together form a supracycle of 15 days since $3 \times 5 = 15$. Within this 15-day cycle, the 3-day cycle will repeat five times, and the 5-day cycle will repeat three times, but any particular pairing of days will occur just once (see Figure 3.1). However, the cycle of pairs for a 4-day cycle combined with a 6-day cycle is only a 12-day supracycle (see Figure 3.2), while combining a 3-day cycle with a 6-day cycle leads only to a cycle of length 6 (see Figure 3.3). The generalization is that the length of the supracycle is given by the *least common multiple* (this is usually abbreviated to l.c.m.), that is, the smallest integer that has each of the component cycle lengths as a factor. For example,

$$\text{l.c.m. } [3,5] = 15, \;\; 3 \times \underline{5} = 15, \;\; 5 \times \underline{3} = 15;$$

$$\text{l.c.m. } [3,6] = 6, \;\; 3 \times \underline{2} = 6, \;\; 6 \times \underline{1} = 6;$$

$$\text{l.c.m. } [4,6] = 12, \;\; 4 \times \underline{3} = 12, \;\; 6 \times \underline{2} = 12$$

where the integer underlined is the number of complete repetitions of the component cycle.

Notice in Figures 3.1 and 3.2 how the component cycles interlock. In the first, the length of the supracycle is the product of the lengths of the two component cycles, while in the second, the length is smaller than their product. And in Figure 3.3, the 3-day component cycle falls

3–day cycles	D1	D2	D3	D1	D2	D3	D1	D2	D3	D1	D2	D3	D1	D2	D3
5–day cycles	d1	d2	d3	d4	d5	d1	d2	d3	d4	d5	d1	d2	d3	d4	d5

Figure 3.1 The 15-day cycle composed of 3-day cycles and 5-day cycles. For pairings, read down each column.

4–day cycles	D1	D2	D3	D4	D1	D2	D3	D4	D1	D2	D3	D4
6–day cycles	d1	d2	d3	d4	d5	d6	d1	d2	d3	d4	d5	d6

Figure 3.2 The 12-day cycle composed of 4-day cycles and 6-day cycles. For pairings, read down each column.

wholly within the 6-day cycle, and so, to contain them both, no additional cycle is created. The difference is that, in the first instance, the component cycle lengths are *relatively prime*; that is, they contain no common factors other than 1, while in the second instance, they are not relatively prime since 2 is a factor of both 4 and 6. In the third case, one component cycle length is a factor of the other; that is, one cycle length is a *divisor* of the other. These principles can be extended to any number of component cycles. For example:

$$\text{l.c.m. } [3, 5, 7] = 105; \text{ l.c.m. } [2, 3, 5, 10] = 30;$$

$$\text{l.c.m. } [3, 4, 6] = 12; \text{ l.c.m. } [3, 6, 12] = 12.$$

In the calendar of the Akan people, most of whom live in Ghana, there is a 6-day week, a 7-day week, and a 42-day supracycle. When counting the number of days in a cycle, the Akan count the same named day as both the first and last day of the cycle. Hence, the 6-day week is

3-day cycles	D1	D2	D3	D1	D2	D3
6-day cycles	d1	d2	d3	d4	d5	d6

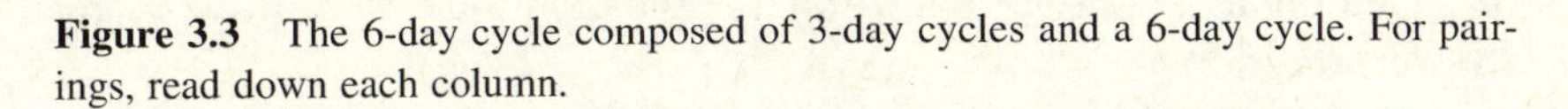

Figure 3.3 The 6-day cycle composed of 3-day cycles and a 6-day cycle. For pairings, read down each column.

referred to as *nanson*, which is literally *seven days*, and the 7-day week is *nawotwe*, literally *eight days*. Although the etymology of the word is unclear to me, the supracycle is referred to as *adaduanan*, literally the *forty days*. Another example is the Northern Thai calendar which has, among other cycles, a cycle of 60 day-names formed by combining a 10-day name cycle with a 12-day name cycle. This cycle is widespread throughout Southeast Asia as it is related to the 60-year name cycle in the Chinese calendar. The 60 year names in the Chinese calendar are formed from a cycle of 10 *celestial stems* with a cycle of 12 *terrestrial branches*, the latter being different types of animals.

We will now examine in much more detail two calendars, each involving several cycles. The calendars are quite dissimilar as are the cultures they are from. To begin, we will look at the calendar of the Maya of Mesoamerica and then turn to the Balinese of Indonesia.

2 The Mayan peoples have a complex cultural tradition extending over millennia and encompassing different groups speaking about 25 different languages. The different groups shared much in the way of culture, but, spread through time and space, they had different centers and political organizations, some different ideas, and some different practices. Scholars now place the beginnings of a distinctive Mayan culture sometime around 1000 BCE. The period 200 CE to 1000 CE is generally referred to as the *Classic period* which is marked by ceremonial centers with monumental architecture, a system of writing, an elaborate astrological science, and numerous centers of social, religious, economic, and political activities. The populations and activities of these centers were interrelated by marriage networks and by trade networks. During the Classic period, the Maya inhabited what are now known as the eastern Mexican states of Chiapas, Tabasco, Campeche, Quintano Roo, and Yucatan. They also extended into Belize, Guatemala, and the western portions of Honduras and El Salvador. Just prior to the Spanish conquest in the sixteenth century, there were, spread out in this area, a number of independent yet culturally interrelated enclaves, none of which were as grandiose as during the earlier Classical period. Because, in part, these Mayan groups were dispersed and independent, they did not succumb to the Spanish as quickly and easily as did some other Amerindians, such as the Incas. Today, primarily in Chiapas and in the highlands of Guatemala, about two million people continue to follow many Mayan traditions.

Christopher Columbus, in 1502, is said to have been the first European to encounter the Maya, and his brother, Bartholomew, was the first to record their name. By mid-century, the Spanish were well established in the area. At that time, remains of the Classic Maya period were already overgrown, and so another "discovery"—this time archeological—took place, beginning in the mid-1800s. Thus, what we know of the Maya, and in particular of their mathematical ideas, comes from several different types of sources. There are, from the Classical period, archeological materials, including thousands of inscribed stone monuments called *stelae*. From the time between 1000 and 1500 CE, there are four books, referred to as *codices*, the only ones remaining of the *thousands* that were burned by the Spanish. And there are reports and descriptions of traditions that continued and of some that are ongoing through this century. Of primary importance is that the Maya had written records kept by specially trained scribes. Their form of writing, however, was modified by the superimposition of Spanish culture, and so the pre-conquest forms require decipherment, which is still ongoing.

Throughout their history, a preoccupation with time pervades the Maya culture. Time is considered to be cyclic. Supernatural forces and beings are associated with, and influence, units of time. Cosmic time and human time are interrelated, and events of the past, present, and future are linked through the recurrence of named time units. There are, indeed, not just one, but several, overlapping cycles, all of which must be taken into consideration to give meaning to any particular time unit. Although their calendric concerns extend to the incorporation of astronomical phenomena, their primary focus was the interrelationship of the arbitrary cycles they created and imposed on time. For this reason, the Maya are said to have "mathematized" time and, through it, their religion and cosmology.

The numbers 13 and 20 are both of great importance in Mayan thought, and so cycles of these lengths are fundamental to their calendar. Each day in each cycle is influenced by a god or, more specifically, *is* a living god. The more concurrent cycles there are, the more gods there are interacting on, and influencing, any particular day. With the 13-day cycle, the days are identified by numbers from 1 to 13, perhaps corresponding to the thirteen sky gods. The gods of 4, 7, 9, and 13, for example, are disposed kindly toward humans, while those of 2, 3, 5, and 10 are malign. Within the 20-day cycle, each day is identified by the name of its god. The sounds of their names differ in the different Maya languages, but they are conventionally transcribed from Yucatec, as

Table 3.1 The gods in the 20-name cycle of the Sacred Round

(D1) Imix	(D11) Chuen
(D2) Ik	(D12) Eb
(D3) Akbal	(D13) Ben
(D4) Kan	(D14) Ix
(D5) Chicchan	(D15) Men
(D6) Cimi	(D16) Cib
(D7) Manik	(D17) Caban
(D8) Lamat	(D18) Eznab
(D9) Muluc	(D19) Cauac
(D10) Oc	(D20) Ahau

shown in Table 3.1. As well as being the god of a particular day, Imix, for example, is god of the earth, Kan is the corn god, Cimi is the god of death, and Ahau is the sun. Since we will be concerned with the place of the god-name in the 20-day cycle, we will instead refer to them as D1, D2, ..., D20.

Because 13 and 20 are relatively prime, the 13-day cycle and 20-day cycle together form a 260-day supracycle. The name of the supracycle, coined by contemporary scholars, is *tolzin*, a Maya word meaning *counting of the days*, or more commonly referred to as the *Sacred Round*. Some scholars have attributed an astronomical or biological basis to the 260-day cycle (such as its closeness to the human gestation period, which is 266 days). Others conclude, and I share their conclusion, that it simply follows from the combination of its component cycles. Each of the 260 days in the Sacred Round will be referred to here as *i*D*j* where *i* is the day number (1–13) and D*j* is the god-name (D1–D20). Day 5D12 (5Eb), for example, is simultaneously the fifth day in the thirteen-number cycle and the twelfth day in the 20-god-name cycle.

There are, however, additional cycles involved in fully identifying a day. Another important cycle contains 365 days and is referred to by some as the *solar year* and by others as the *Vague Year*. I prefer Vague Year because, while the number of days may have been inspired by the solar cycle, it became a cycle of fixed length without adjustments to keep it in synchronization with the sun. The 365-day cycle contains a 360-day supracycle made up of a cycle of 20 numbers and a cycle of 18 named gods, plus 5 residual days.

The relationships of the 20 numbers and 18 names in this 360-day supracycle differ from the relationship of the 13 numbers and 20 names

3-number cycle	1	2	3	1	2	3	1	2	3	1	2	3	1	2	3
5-name cycle	A1	A1	A1	A2	A2	A2	A3	A3	A3	A4	A4	A4	A5	A5	A5

Figure 3.4 A 3-number (1, 2, 3) cycle *within* a 5-name cycle (A1, A2, A3, A4, A5). For pairings, read down each column. (Compare this with Figure 3.1.)

that constitute the 260-day supracycle of the Sacred Round. As with the days and months in the Gregorian calendar, where a full set of day numbers is used before the month name advances, the 20-number cycle is *within* the cycle of 18 names. If there were, for example, only three numbers *within* five god names, an overall supracycle of length 15 would result as shown in Figure 3.4. When one cycle is within another, the length of their supracycle is always the product of their lengths.

For the cycle of 18 gods in the Vague Year, the names are shown in Table 3.2. Here again, to keep our focus on the order of the gods within the cycle, we will refer to them as Ak where $k = 1, \ldots, 18$. Thus, each day within the Vague Year, except for the last five residual days, would be mAk where m is the day number (1–20) and Ak is the god-name (A1–A18). (The day number *20* was not actually used. Instead, there was sometimes a figure that stood for *end* or *final* and, at other times, a figure that referred to the *seating* of the next month. In transcriptions, the latter is conventionally designated as *0* with the name of the next month. To avoid implying the use of the numeral *zero*, and to keep our focus on the order of the days in the cycle, we have referred to the last day number as *20*. The residual days were labeled 1, 2, 3, 4, and, again, a figure indicating *end*, which we will call *5*.)

The Vague Year (365 days) and the Sacred Round (260 days) are combined by the Maya into the *Calendar Round*, a still larger cycle incorporating them both. Its length is l.c.m. [260,365] = 18,980 days, which is equivalent to 52 Vague Years or 78 Sacred Rounds. Each day within this larger cycle is identified by adjoining its subcycle identifiers, namely, in our notation as iDj, mAk where $i = 1, \ldots, 13$; $j = 1, \ldots, 20$; $m = 1, \ldots, 20$; $k = 1, \ldots, 18$. There are, however, also the residual days. Thus, for example, looking at the 14 days near the end of one arbitrary Vague Year and the beginning of the next, the days would be:

Table 3.2 The gods in the 18-name cycle of the vague year

(A1) Pop	(A10) Yax
(A2) Uo	(A11) Zac
(A3) Zip	(A12) Ceh
(A4) Zotz	(A13) Mac
(A5) Tzec	(A14) Kankin
(A6) Xul	(A15) Muan
(A7) Yaxkin	(A16) Pax
(A8) Mol	(A17) Kayab
(A9) Chen	(A18) Cumku

Last 8 days of an arbitrary Vague Year

1D4, 18A18
2D5, 19A18
3D6, 20A18
4D7, 1
5D8, 2
6D9, 3
7D10, 4
8D11, 5

First 6 days of the next Vague Year

9D12, 1A1
10D13, 2A1
11D14, 3A1
12D15, 4A1
13D16, 5A1
1D17, 6A1

In the ceremonial centers of the Classic period, there were hundreds of stelae erected to commemorate different events. To mark an event, in addition to a Calendar Round date, another significant identifier was a *Long Count* date. The Long Count date depended on yet *another* cycle—the *Great Cycle*. A Great Cycle is based on a 360-day period (a *tun*) consisting of 18 *uinals* of 20 days each. There is some speculation that this 360-day period arose as an early version of the solar year or of the Vague Year. But, whether or not it did, the tun became a distinct arbitrary cycle in and of itself. Twenty tuns are a *katun*; 20 katuns are a *baktun*; and 13 baktuns are a Great Cycle. A Long Count date is the number of days from the beginning of the then ongoing Great Cycle.

An example of a Long Count date, transcribed into our numerals, is 9.0.19.2.4. From left to right, this reads "9 baktuns, 0 katuns, 19 tuns, 2 uinals, and 4 days." To convert to our decimal system, starting at the right, each position—with the exception of the third—is multiplied by one higher power of 20. In the third position, an 18 is used instead. Hence, the Classical Long Count date of 9.0.19.2.4 is interpreted as:

$$9 \cdot 18 \cdot 20^3 + 0 \cdot 18 \cdot 20^2 + 19 \cdot 18 \cdot 20 + 2 \cdot 20 + 4$$
$$= 1{,}302{,}884 \text{ days}$$

from the beginning of the Great Cycle which started on the Calendar Round date of 4D20, 8A18 (4 Ahau 8 Cumku).

What is most significant about a Great Cycle is that it is anchored at a starting point and, so, enables the pinpointing of historical events and the measurement of elapsed time. As such, it adds measures that are usually associated with linearity. In mathematical terms, we could say that just as any segment of the circumference of a circle of infinite radius can be thought of as a straight line, a Great Cycle (consisting of 1,872,000 days) is a cycle so large that it enables linear-like measures.

Numbers of elapsed days, as well as dates, appeared on stelae in Long Count form. When used as statements of elapsed days, the numbers can take on values greater than the number of days in a single Great Cycle. In Long Count *numbers*, just as 20 tuns make up a katun and 20 katuns make up a baktun, so, in general, each additional position introduces another multiple of 20. These Long Count numbers are associated with Maya computations that are projections into the past or into the future and which dovetail the different calendric cycles. For instance, one inscription, commemorating the enthronement of a ruler, gives the Calendar Round dates of his birth and his enthronement, as well as of the enthronement of an earlier, somehow related, ruler or deity. The number of days elapsed since the enthronement of the deity, given in Long Count form, is 7.18.2.9.2.12.1 days. Hence, given one Calendar Round date, a Calendar Round date some 1.25 million years (455,393,761 days) earlier was calculated, or, given two Calendar Round dates, their Long Count difference was calculated.

There were other cycles used on the stelae to mark events, and numerous other complex calculations, particularly in the Mayan codices of the eleventh century CE. However, we will remain focused on the Classical period and further examine the mathematical ideas evidenced by the stelae and the cycles already introduced. Just these cycles provide ample evidence that the Mayan calendrical ideas intimately related human affairs with temporal cycles, and the interplay of the cycles was of paramount importance. When dealing with cycles, however, the mathematical ideas involved in calculations of elapsed time need further elaboration to be fully appreciated.

Because the stelae are marked by both Calendar Round dates (which are made up of Sacred Round and Vague Year dates) and Long Count dates, there are several aspects to elapsed time calculations. They are:

A. Adding (or subtracting) *N* days to a Calendar Round date:
   1. Adding *N* days to a date in the Sacred Round;
   2. Adding *N* days to a date in the Vague Year;
   3. Subtracting *N* days from a date in the Sacred Round or Vague Year;

B. Adding (or subtracting) *N* days to a Long Count date;

C. Finding the number of days, *N*, between two Long Count dates; and

D. Finding the number of days, *N*, between two Calendar Round dates.

As we examine these, in turn, it is important to keep in mind that although dates include numbers, dates and numbers are different kinds of things. In the more familiar Gregorian system, for example, 9/4/91 is a date, while 211 is a number. None the less, they can be combined; adding 211 days to the date 9/4/91 results in another date (4/29/92). The calculations will, of necessity, contain detailed steps and even some theorems. If you are wary of these, you may scan the rest of this section and still be able to fully rejoin us in Section 3.

A. Addition (or subtraction) of N days to a Calendar Round date

1. To add *N* days to a Sacred Round date $i\mathrm{D}j$ ($i = 1, \ldots, 13$; $j = 1, \ldots, 20$), the cycles that constitute it are advanced independently. That is,

$$i\mathrm{D}j+N \text{ days} = (i + N) \text{ (mod 13) } \mathrm{D}(j + N).$$

For example:

$$5\mathrm{D}7 + 97 \text{ days} = 102 \text{ (mod 13) } \mathrm{D}104 \text{ (mod 20)} = 11\mathrm{D}4.$$

$$5\mathrm{D}7 + 16 \text{ days} = 21 \text{ (mod 13) } \mathrm{D}23 \text{ (mod 20)} = 8\mathrm{D}3.$$

2. Adding N days to a vague year date $m\text{A}k$ ($m = 1, \ldots, 20$; $k = 1, \ldots, 18$) proceeds differently because, rather than being independent, one of the two cycles is inside the other and, in addition, there are the five residual days. The date is first converted to its number of days from the start of the year, and then N is added to that number. (This is analogous to adding, say, 40 days to February 15 by first converting February 15 to 46 days from the start of the year and then adding 40.) The result,

$$N' = [20\ (k - 1) + m + N] \bmod 365,$$

must then be reconverted to date form. If N' is less than or equal to 360, the resulting date $m'\text{A}k'$ is:

$$m' = N'\ (\text{mod } 20) \text{ and } k' = (N' - m' + 20)/20.$$

If N' is greater than 360, the excess above 360 is the number of the residual day.

For example:

$$4\text{A}10 + 97 \text{ days} : N' = (180 + 4 + 97)\ (\text{mod } 365) = 281;$$

$$m' = 281\ (\text{mod } 20) = 1\ ;$$

$$k' = (281 - 1 + 20)/20 = 15;$$

Result : 1A15.

$$4\text{A}10 + 543 \text{ days} : N' = (180 + 4 + 543)\ (\text{mod } 365) = 362;$$

Result: residual day 2.

3. Subtractions of N days from a Sacred Round date or from a Vague Year date proceed, as do the additions of N days just described in A.1 and A.2. They involve, however, an additional modular relationship, namely

$$-a\ (\text{mod } b) = (b - a)\ (\text{mod } b).$$

That is, for example, $-5\ (\text{mod } 7) = 2\ (\text{mod } 7)$.

Examples of subtractions are:

$$5\text{D}7 - 97 = -92\ (\text{mod } 13)\ \text{D}\ (-90) \bmod 20$$

$$= -1\ (\text{mod } 13)\ \text{D}\ (-10) \bmod 20 = 12\text{D}10.$$

$$4\text{A}10 - 196 = -12\ (\text{mod } 365) = 353 \text{ resulting in } 13\text{A}18.$$

The order and form of these calculations are mine. No one knows how the Maya actually did them. Notice, for example, that here, the arithmetic is being done in the decimal system rather than in the Maya base twenty system, and N was assumed to be a decimal number rather than in Long Count difference form. If N were in Long Count form, we could first convert it to decimal form and then proceed as above. The conversion would be done as follows:

$$\text{Long Count } a_5.a_4.a_3.a_2.a_1 = ([(a_5 \cdot 20 + a_4)20 + a_3]\ 18 + a_2)\ 20 + a_1 \text{ decimal.}$$

The Long Count number form was used by the Maya only for calendrics and is not, in general, the form of Maya numbers. Maya numbers are strictly base twenty; that is, each consecutive position is valued at one higher power of twenty, with no intrusive value of eighteen involved.

B. Adding (or subtracting) a Long Count number of days, N, to a Long Count date

In this form, in each position, except for the second position from the right, numbers can go as high as 19. When adding, if the sum of two numbers in a position is 20 or greater, the amount above 20 is recorded, and a one is carried to the next column. In the second position, the number can only go as high as 17. When the sum is greater than or equal to 18, the excess above 18 is recorded and a one carried to the next column.

Example :						
	5.	3.	9.	15.	1	(date)
	+6.	12.	14.	10.	3	(number)
	11.	16.	4.	7.	4	(date)

Subtraction is similar except that one borrows from an adjacent column rather than adding a carry to it.

Example :						
	7.	8.	3.	14.	2	(date)
	−6.	12.	14.	10.	3	(number)
		15.	9.	3.	19	(date)

Example :	7.	8.	12.	3.	5	(date)
	−3.	9.	11.	10.	4	(number)
	3.	19.	0.	11.	1	(date)

C. Finding the Long Count number of days, N, between two Long Count dates
This calculation proceeds exactly as did the subtraction just above. The only difference is that a date is being subtracted from a date, and the result is a number.

D. Finding the number of days, N, between two Calendar Round dates
Once again, we use decimal numbers and decimal arithmetic as we did when working with Calendar Round dates in part A. Because the dates are composites of several cycles and are not anchored to a single temporal starting point, they again involve the arithmetic of cycles, which is conceptually different from, and, for us, more difficult than the arithmetic of linear measures. Suppose that the problem is to find N, the number of days between 12D8, 6A2 and 5D4, 12A17 assuming that the latter comes later. (Such an assumption must be added because each Calendar Round date reoccurs every 18,980 days, and so each of the two dates is both before and after the other!) For the Sacred Round portion of the dates, an N is needed, which simultaneously satisfies the difference in the day numbers and the difference in the god-names. That is, it must satisfy both $N = (5-12) \bmod 13$ and $N = (4-8) \bmod 20$. And, for the Vague Year portion, an N is needed satisfying the difference of the numbers of days in that cycle: $N = [(17-2)\,20 + (12-6)] \bmod 365$. Thus, simplifying these, a single N must *simultaneously* satisfy the three conditions

$$N = 6 \pmod{13},\ N = 16 \pmod{20},\ \text{and } N = 306 \pmod{365}.$$

For each condition individually, there are an infinite number of solutions:

$$N = 6,\ 19,\ 32,\ 45,\ \ldots \text{ for the first;}$$

$$N = 16,\ 36,\ 56,\ 76,\ \ldots \text{ for the second; and}$$

$$N = 306,\ 671,\ 1036,\ 1401,\ \ldots \text{ for the third.}$$

The simultaneous solution must be a number appearing on all three of these lists. There are, however, an infinite number of such numbers, just as there are an infinite number of answers to the question, "How many months beyond April is May?" The answers could be 1, 13, 25, 37, and so on; that is, $1 + 12Q$ where Q is any integer. The question can be made more precise by specifying the solution to be the smallest number, but none the less, finding that smallest N is still a difficult problem.

In modern mathematics, the problem is deemed "solving a system of linear congruences expressible as $x = a_i \pmod{m_i}$ where $i = 1, \ldots, n$." Its modern solution is attributed to Carl F. Gauss, a German mathematician, in 1801, but it has long been recognized that a complete method of solution existed in China in the thirteenth century. In fact, many modern mathematics texts refer to the problem as "The Chinese Remainder Problem," or even to its modern solution as "The Chinese Remainder Theorem." Further, the Chinese work containing the solution included calendric problems *of the type being discussed here*. Additionally, there are statements of such problems in both China and India as early as the fifth century. Thus, we see that while these cultures had calendars different from each other, and different from the Maya, all shared a concern for concurrent cycles and posed questions and sought solutions to problems associated with them. We do not know how the Maya solved these problems. Nevertheless, we do know that they did solve them. And we also know that within the scribal class, there were highly trained specialists who dealt with issues relating to calendrics.

The *modern* mathematical solution will be detailed here, although it is surely not the way the Maya handled the problem. In our example above, the problem was recast to finding N satisfying:

$$N = 16 \pmod{20} = 6 \pmod{13} = 306 \pmod{365}.$$

A difficulty is that 20 and 365 share a common factor, and so we must begin by rewriting them in a form that eliminates that commonality. In general, for b and c with no shared factors:

$$a \pmod{bc} = a \pmod{b} = a \pmod{c}.$$

Hence, since $20 = 4{\cdot}5$ and $365 = 73{\cdot}5$,

$$16 \pmod{20} = 16 \pmod{4} = 0 \pmod{4} = 16 \pmod{5}$$
$$= 1 \pmod{5},$$

and

$$306 \pmod{365} = 306 \pmod{73}$$
$$= 14 \pmod{73} = 306 \pmod{5} = 1 \pmod{5}.$$

The problem, therefore, can be restated as to find N satisfying:

$$N = 0 \pmod 4 = 1 \pmod 5 = 6 \pmod{13} = 14 \pmod{73}.$$

The difficulty of common factors has been eliminated, but in so doing, the number of simultaneous conditions was increased.

Step 1:

$$N = 0 \pmod 4 = 0 + 4p \text{ for some integer } p. \tag{1}$$

Also,

$$N = 1 \pmod 5 \text{ so } 4p = 1 \pmod 5. \tag{2}$$

Step 2: To solve (2) for p, Euler's theorem can be used. It states that for k and m relatively prime, if $kp = c \pmod m$, then $p = ck^{\varphi(m)-1}$ (mod m) where $\varphi(m)$ is the number of integers less than or equal to m that are relatively prime to m. In this case, that means $p = 1\ (4)^{\varphi(5)-1} \pmod 5$. Since the integers less than or equal to 5 and relatively prime to 5 are 1, 2, 3, 4, $\varphi(5) = 4$.

Hence,

$$p = 1(4)^3 \pmod 5 = 64 \pmod 5 = 4 \pmod 5$$

$$= 4 + 5b \text{ for some integer } b. \tag{3}$$

[Alternately, if possible, we can solve by observation. For $4p = 1$ (mod 5), we observe that 1 (mod 5) = 16 (mod 5) and so easily solve $4p = 16$ (mod 5) which gives $p = 4 \pmod 5$, as does Euler's theorem.]

Step 3: Combining (1) and (3),

$$N = 4p = 4\ (4 + 5b) = 16 + 20b. \tag{4}$$

Combining (4) with $N = 6 \pmod{13}$,

$$6 \pmod{13} = 16 + 20b$$

and so

$$-10 \pmod{13} = 3 \pmod{13} = 20b. \tag{5}$$

Step 4: To solve (5) for b, once again, Euler's theorem is used. Here, the result is

$$b = 3\ (20)^{\varphi(13)-1} \pmod{13} \text{ where } \varphi(13) = 12,$$

$$b = 3\ (20)^{11} \pmod{13} = 6 \pmod{13} = 6 + 13c \text{ for some integer } c. \tag{6}$$

Step 5: Combining (4) and (6)

$$N = 16 + 20b = 16 + 20\ (6 + 13c). \tag{7}$$

Combining (7) with $N = 14 \pmod{73}$,

$$14 \pmod{73} = 136 + 260c$$

and so

$$-\ 122 \pmod{73} = 24 \pmod{73} = 260c. \tag{8}$$

Step 6: To solve (8) for c, we will use observation instead of Euler's theorem. For $260c = 24 \pmod{73}$, observe that $24 \pmod{73} = 14{,}040 \pmod{73}$ and so solve $260c = 14{,}040 \pmod{73}$, which gives $c = 54$.

Step 7: Substituting $c = 54$ into (8), $N = 14176$.

From the modern formulation of the solution of simultaneous linear congruences, an interesting insight is gained into the structure of the Mayan calendar. One modern theorem states that the problem

$$x = a_i \pmod{m_i}\ i = 1,\ \ldots,\ n$$

has a solution *if and only if* for each pair i, j, where $1 \leq i < j < n$, the greatest common divisor of m_i and m_j evenly divides $(a_i - aj)$. This tells us that, for example, the problem of finding the number of days between 4D20, 8A18 and 12D12, 8A11 has *no* solution. That is, both cannot be valid dates in the Mayan calendar! Because we know 4D20, 8A18 to be a valid date, namely the start date of a Great Cycle, the date 12D12, 8A11 must be invalid. Clearly, when assigning dates to events in the past or future, the Maya would have had to have some way to assure that only valid dates were created.

3 This discussion gives only a taste of the Maya involvement with calendrics. Leaving aside the considerable evidence from the codices that links calendars, mathematical ideas, and astronomical knowledge, we now continue with our primary concern in this chapter—the widespread, but diverse, use of arbitrary abstract cycles in the creation of calendars and in conceptions of time. We move to another area of the world and another culture. In our next example, a calendar from Bali, there are also cycles within cycles and cycles superimposed on cycles, but just as the cultures are different, juxtaposition of the cycles is different, the surrounding beliefs are different, and the associated mathematical ideas are different.

4 The island of Bali is just east of Java, separated from it by a narrow strait (see Map 1 in Chapter 2). Both are part of Indonesia, which consists of about 13,000 islands spread for 6400 km along the imaginary line that we call the equator. About 6000 of the islands are named and about 1000 permanently settled. The overall population of Indonesia is about 180 million people. Of these, some two million live on Bali. Indonesia has been one nation since 1949, but its people and their cultures and histories are quite diverse. The Kodi, whose calendar was discussed in the previous chapter, also live in Indonesia, on the island of Sumba some 500 km west of Bali; their culture and their calendar are decidedly different.

The island of Bali has an area of about 5000 sq. km. Mountains as high as 3140 m on the northern side of the island slope down to the sea on its southern side. For more than 1000 years, and still during the twentieth century, the majority of the people were engaged in wet rice cultivation, living in villages made up of compounds enclosed by clay walls. Bamboo, coconut, and banana trees are planted within the compounds, while outside of the walls are the surrounding wet rice fields. The social and religious organizations and activities of the villages are intertwined with the planting and irrigating of the fields. A primary function of the groups and their activities is to assure the well-being of the inhabitants by maintaining cosmic balance. So doing involves the people in the worship of divine powers via the care and maintenance of the temples, the giving of offerings, and by participation in festivals and ceremonies.

Balinese culture is now a blend of a variety of influences. Speculation on the place of origin of the modern indigenous population includes Taiwan, off the east coast of China; Southwest China; and India; all prior to 1000 BCE. There is evidence of early trade with other islands and with mainland Southeast Asia. There was decided influence of Indian culture and the Hindu religion prior to 1000 CE, and of Javanese culture and Islam in about 1200 CE. Then, beginning in the sixteenth century, with the coming of the Portuguese and then the Dutch, there was European and Christian influence. As a result of the blending, there are now a number of distinct calendars existing simultaneously on Bali with some events scheduled or measured in one and some in another.

Here, we are concerned only with the calendar called the Javanese–Balinese calendar. As contrasted to other calendars on Bali, this calendar involves no lunar or solar cycles, and its purpose is not the measurement of elapsed time.

In Balinese cosmology, the world in which people live—the Middle World—is between the Upper World of the gods and the Lower World of the demons. The Upper and Lower worlds control cycles of growth and decay, which intersect in the Middle World, resulting in the life cycles and life processes of all living things. Thus, nature, the physical universe, and the lives of human beings, all of which occur in the Middle World, are shaped by forces outside of themselves, and those forces are cyclic. Hence, the structure of the calendar reflects that each day is part of many cycles, and the purpose of the calendar is to properly present the multiplicity of intersections that characterize each day. Where a day falls in each of the cycles individually and in the cycles in combination determine what should be done or avoided, whether it is lucky or unlucky, which gods are involved, which obligations are required, or, in short, the personal, religious, and cultural significance of the day.

The Javanese–Balinese calendar contains ten different arbitrary cycles, which are usually referred to as ten different length weeks; there is a 10-day week, a 9-day week, an 8-day week, a 7-day week, and so on, down to a 1-day week. A year in this calendar, that is, a full supracycle of all of these, has 210 days. There are evenly within the year twenty-one 10-day weeks, thirty 7-day weeks, thirty-five 6-day weeks, forty-two 5-day weeks, seventy 3-day weeks, one hundred and five 2-day weeks, and two hundred and ten 1-day weeks. That is, $210 = 1 \cdot 2 \cdot 3 \cdot 5 \cdot 7$ and so is divisible by these factors and their products. In order to fit into the year, special adjustments are used for weeks of length 4, 8, and 9. Just as leap days, or even leap months, are added periodically into calendars that were discussed in the previous chapter, here some specific days are repeated in particular weeks once each year. Were it not for these adjustments, a grand cycle containing all the cycles evenly would have to be far larger than 210 days; its length would have to be divisible by 8, 9, 5, and 7 rather than just by 2, 3, 5, and 7, and so it would be 2520 days.

Just as every day in the Gregorian 7-day week has a different name, every day in *each* of the Balinese weeks has a different name. Hence, each day has 10 different names. In all, since there are ten names in the 10-day cycle, nine in the 9-day cycle, and so on, there are $10 + 9 + 8 + 7 + \cdots + 1 = 55$ different day-names. For simplicity, and to highlight their cyclic positions, we will use the names i_n where n is the week length, and, starting with some arbitrary day, i is the day of the cycle, where $i = 1, \ldots, n$. For example, for the Gregorian week, arbitrarily starting with Sunday, Wednesday would be 4_7; that is, the 4th day of a 7-day cycle. Since each of the days in the Javanese–Balinese year has ten names, to be fully identified, a day must have ten such identifiers—for example (1_1, 2_2, 3_3, 1_4, 1_5, 6_6, 3_7, 1_8, 3_9, 6_{10}).

In addition to determining major cultural festivals and ceremonies and activities particular to each temple, the calendar is intimately linked to the personal identity of each individual. For example, the day within the 8-day cycle on which a child is born is a clue to the child's identity in a former incarnation. The day of birth in the 5-day and 7-day cycles determines the offerings that need to be made to pay the child's debts to beings in the spirit world. Throughout the child's first year, there are important ceremonies on auspicious days. For the first 42 days (that is, six 7-day weeks or seven 6-day weeks), the mother and child are considered impure. The ceremony on the 42nd day removes the impurities. For the first 105 days (that is, three 35-day supracycles), the child is vulnerable to witches or sorcerers, and the ceremony on the 105th day protects the child and increases its strength. At the end of its first year, that is, on the 210th day, the child's hair is first cut. Only after that may the child's feet touch the ground; until then, it is carried wherever it goes. Further, identifying a person by place within cycles is also seen in birth-order and generational-order cycles. Each child is named according to a four-name cycle so that a family's fifth child is given the same name as the first, the sixth as the second, and so on. In kin terminology, four generations are named with the fifth returning to the same name as the first. As a result, for example, when making offerings at a family's ancestral shrine, a child is told not to pray to the parents' great-grandparents as they are in the same generation as is the child.

In general, extreme importance is attached to where a day falls in the conjunction of the 5-day and 7-day weeks. The 35-day supracycle containing all of their intersections is often pictorially represented. The representation, called a *plintangan* calendar, contains a five-row by seven-column array of symbolic scenes bordered above by gods, trees, and birds, and below by demons and animals. The resulting configuration is, therefore, seven rows by seven columns, also sometimes seen as separate representations containing individual columns of seven rows each.

Another important calendric artifact, called a *tika*, represents the structure of the year and marks particularly important communal days within it. The tika are colorful objects; some found in museums

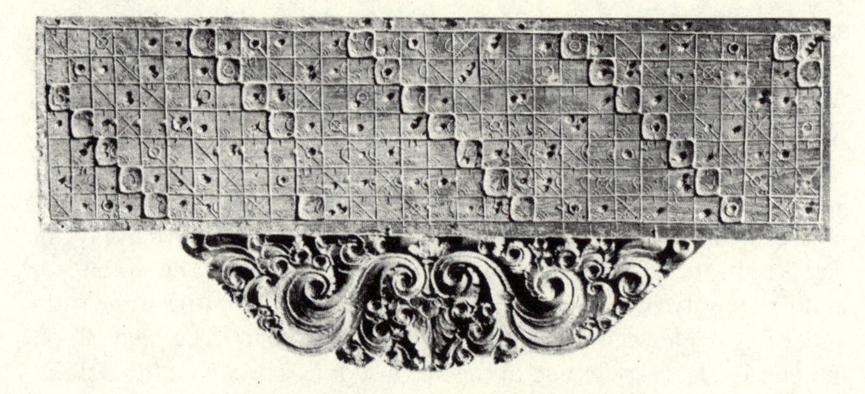

Figure 3.5 A carved wooden tika.

are painted or carved on wood, and those now commonly available are printed on paper. Although, perhaps, in earlier times, their ownership was restricted to specialists in calendric interpretation, tika are now sold in public markets.

The tika is arranged as a rectangle of seven rows and thirty columns. Thus, it emphasizes the 7-day week and also highlights that there are 30 such weeks. Each 7-day week is called a *wuku*, and together, the 30 weeks are referred to as the *wuku year*. And just as each of the 7 days has a unique name, each wuku has a different name. Hence, knowing the day in the 7-day week (row) and also the wuku in the 30-week cycle (column), a day can be located on the tika, and where the day stands in every cycle can then be read from it. In order for us to read the tika, however, the symbols used on it must be examined more closely. Here, two tika are shown: a wooden one, at least 100 years old, from a museum collection (Figure 3.5); and a common paper version (Figure 3.6).

Figure 3.6, which we examine first, has six different symbols, each used to mark a particular day in one of the cycles. Figure 3.7 shows the numerical identifiers we associate with each symbol. It is important, however, to keep in mind that although the days are ordered within the

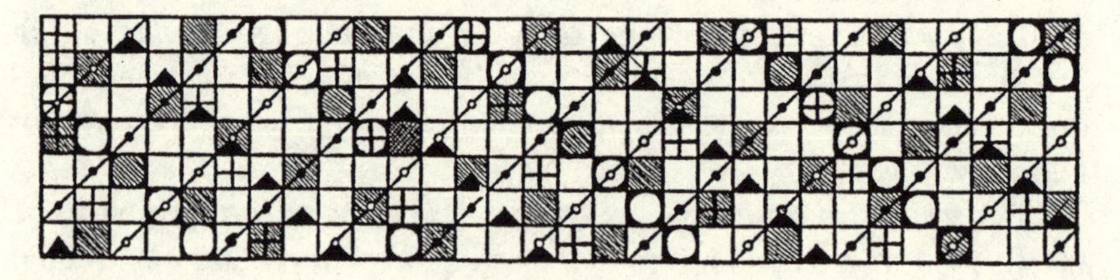

Figure 3.6 A paper tika.

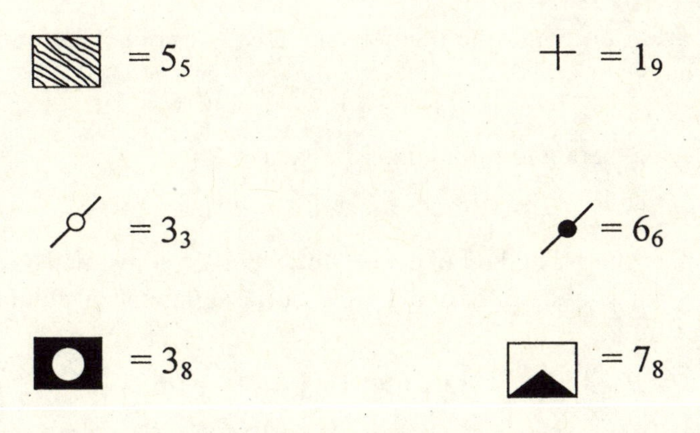

Figure 3.7 Symbols on tika in Figure 3.6 with numerical identifier equivalents. (Note: each 3_8 and 7_8 is also a 3_4; each 6_6 is also a 3_3.)

cycles, the cycles are continuous, and it is we, not the Balinese, who are highlighting the structure of the calendar by using numerical identifiers in place of the 55 day-names.

Since two 4-day cycles are evenly contained within an 8-day cycle, and two 2-day cycles are within a 4-day cycle, we observe that

$$i_8 = i \text{ (mod 4)}_4 \text{ and } i_4 = i \text{ (mod 2)}_2. \tag{9}$$

Four and eight, however, do not evenly divide 210. To make these cycles conform to the year, at a particular point in the year, there are three consecutive days called 7_8. That is, since 210 (mod 8) = 2, two days called 7_8 are added to the twenty-six complete 8-day cycles. Similarly, 210 (mod 4) = 2, and so two additional days are also needed to reconcile the 4-day weeks with the year. This, too, is taken care of by the two inserted days just described, since days called 7_8 are also called 3_4. As a result, despite the insertion, the alignment of names in the 2-, 4-, and 8-day cycles remains unchanged!

Another cycle that does not fit neatly is the 9-day cycle: 210 (mod 9) = 3. For that, a similar insertion is made but at a different point in the year. At that point, day-name 1_9 is repeated consecutively four times. The 3-day week names, however, remain unmodified, and so the three inserted days are an exception to the 3-day and 9-day name cycle alignment, while the 3-day and 6-day alignment remains unchanged as do the 2-day and 6-day alignment. That is, although, in general throughout the year,

$$i_9 = i \text{ (mod 3)}_3, \tag{10}$$

at this insertion point, the relationship does not hold. The four consecutive days are

$$1_9 = 1_3, \quad 1_9 = 2_3, \quad 1_9 = 3_3, \quad \text{and } 1_9 = 1_3.$$

However, here and throughout the year,

$$i_6 = i \ (\text{mod } 3)_3 \text{ and } i_6 = i \ (\text{mod } 2)_2. \qquad (11)$$

Another interrelated set of cycles, for which no adjustment is needed, are those of lengths 2, 5, and 10. By our assignment of numbers, the relationship is:

$$i_{10} = i \ (\text{mod } 5)_5 \text{ and } i_{10} = i + 1 \ (\text{mod } 2)_2. \qquad (12)$$

Our example of a reading of the tika begins with row 7, column 12, because several symbols appear close together. Row 7 indicates that the 7-day week name is 7_7, and the symbols in the square further identify the day as 5_5 and 6_6. The symbols in the next square (row 1, column 13) read 3_8 and 1_9, and so the square being read is 2_8 and 9_9. Further, we know from relationships (9), (10), (11), and (12) that a 6_6 is also a 3_3, a 2_8 is also a 2_4 and a 2_2, and a $(5_5, 2_2)$ combination is also a 5_{10}. Hence, in all, the day is $(1_1, 2_2, 3_3, 2_4, 5_5, 6_6, 7_7, 2_8, 9_9, 5_{10})$.

As another example, let us read row 6, column 25 of the tika. The sixth row indicates that it is 6_7, and the symbols in the square further identify it as 5_5 and 6_6. Just before it is the symbol for 3_8, and just after it is the symbol for 1_9. Hence, in all, the day is $(1_1, 2_2, 3_3, 4_4, 5_5, 6_6, 6_7, 4_8, 9_9, 5_{10})$. Our reading in this example can be verified by addition. The day is 12 columns and 6 rows, that is, $12 \cdot 7 + 6 = 90$ days beyond the previous example. Adding 90 to each item of the first reading indeed gives the result in the second reading:

$$2_2 + 90 \ (\text{mod } 2) = 2_2; \quad 6_6 + 90 \ (\text{mod } 6) = 6_6;$$

$$3_3 + 90 \ (\text{mod } 3) = 3_3; \quad 7_7 + 90 \ (\text{mod } 7) = 6_7;$$

$$2_4 + 90 \ (\text{mod } 4) = 4_4; \quad 2_8 + 90 \ (\text{mod } 8) = 4_8;$$

$$5_5 + 90 \ (\text{mod } 5) = 5_5; \quad 9_9 + 90 \ (\text{mod } 9) = 9_9;$$

$$5_{10} + 90 \ (\text{mod } 10) = 5_{10}.$$

Had the time interval included any of the multiple days, the addition would have had to take that into consideration. The consecutive four days named 1_9 come at the start of the year (see column 1, rows 1, 2, 3, 4 on the tika). The consecutive days named 7_8 and 3_4 occur at an important time of year when, for 2 weeks, there is a sequence of holidays surrounding the coming to earth of the spirits of the ancestors, their residence in the chapels, and then their return to heaven (see column 11, rows 1, 2, 3).

On the tika, one can also see recurring patterns of overlapping symbols, which are concurrences of specific days within different cycles. For example, the concurrence of 5_5 and 3_3, which occurs every 15 days, is propitious for offerings to evil spirits, while the concurrence of 5_5 and 7_7, falling every 35 days, is extremely lucky, as are (5_5, 3_7) and (4_5, 4_7).

The wooden tika in Figure 3.5 has symbolic markings for days in addition to those marked on the tika of Figure 3.6. Even for the same days, however, the markings differ. Figure 3.8 contains a key to the symbols for the 6 days common to both as well as the marking for one additional day. The wooden tika testifies to the importance of these artifacts as well as to the fact that tika, in general, are representations of a shared conceptual model.

In order to uniquely identify any day, the Balinese often give the three names of the day in the 5-, 6-, and 7-day weeks. These are indeed sufficient because, together, they span the 210-day cycle; that is, the place in one of these weeks does not depend on, or

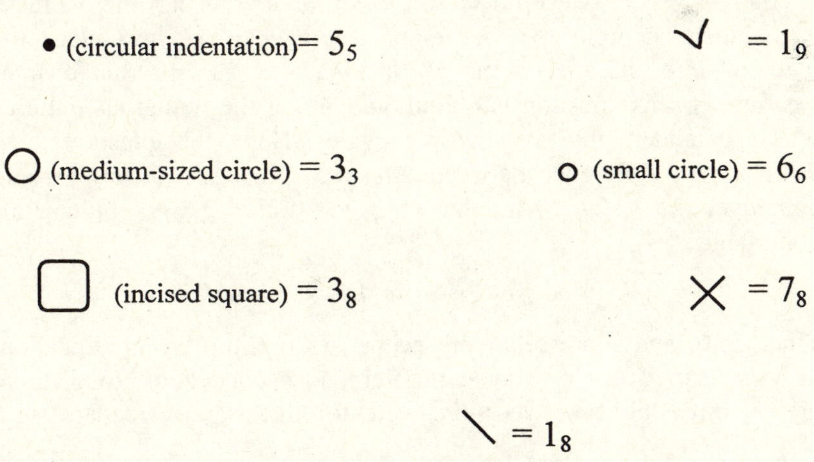

• (circular indentation)= 5_5

√ = 1_9

○ (medium-sized circle) = 3_3

o (small circle) = 6_6

□ (incised square) = 3_8

X = 7_8

\ = 1_8

Figure 3.8 Symbols on tika in Figure 3.5 with numerical identifier equivalents. (Note: each 3_8 and 7_8 is also a 3_4; each 6_6 is also a 3_3.)

determine, the place in another, and taken together, they include all possible days. With the 5-, 6-, 7-day cycle information, the specific day can be readily read from the tika. For example, consider the day (2_5, 3_6, 5_7). Because the day is 5_7, it is in row 5 and, by virtue of it being 2_5 and 3_6, we seek a column in which the symbol for 5_5 is two squares above (row 3) and the symbol for 6_6 is just above that (row 2). There is just one such column—column 29—and so, in all, we can read that the day is,

$$(1_1,\ 1_2,\ 3_3,\ 3_4,\ 2_5,\ 3_6,\ 5_7,\ 7_8,\ 8_9,\ 2_{10}).$$

The Balinese, according to one observer, also readily solve in their heads such problems as: "How many days to Galungan?" [Galungan is a holiday that falls on (5_5, 2_6, 4_7)]. A typical answer is "eighty days from the next Kliwon." (Kliwon is the day-name equivalent to 5_5.) For us, this would be extremely difficult, if at all possible. It becomes straightforward, however, were we to use a tika or a mental image of one. Suppose that the current day, the day on which the question is being asked, is the day (2_5, 3_6, 5_7) discussed in the example above. Putting our finger on that day on the tika—row 5, column 29—we then move forward to the next Kliwon, that is, to the next 5_5, which is in row 1, column 30. Galungan (5_5, 2_6, 4_7) is in row 4, column 11. The difference between row 4, column 11 and row 1, column 30 is 11 columns plus 3 rows, which equals $11\ (7) + 3 = 80$ days.

Alternatively, we could find the difference using the Chinese Remainder Theorem. As was emphasized when we introduced the theorem into the discussion of the Maya earlier in this chapter (section 2), this question is ubiquitous among the numerous cultures whose calendars involve multiple cycles. Here, the question is to calculate the difference between "the next Kliwon" (5_5, 6_6, 1_7) and Galungan (5_5, 2_6, 4_7); that is, to solve the simultaneous congruences,

$$0 \pmod 5) = -4 \pmod 6) = 3 \pmod 7).$$

Since 5, 6, and 7 are relatively prime, no preliminary modification, as was needed in the Mayan problem, is needed here. Also, there are no infeasible dates as there were for the Maya calendar.

Step 1:

$$N = 0 \text{ (mod 5)} = 0 + 5p \text{ for some integer } p. \qquad (13)$$

Also,

$$N = -4 \text{ (mod 6)} = 2\text{(mod 6) so } 5p = 2 \text{ (mod 6)}. \qquad (14)$$

Step 2: To solve (14) for *p*, Euler's theorem can be used. In this case, that gives,

$$p = 2\ (5)^{\varphi(6)-1} \text{ (mod 6) where } \varphi(6) = 2,$$

$$p = 10 \text{ (mod 6)} = 4 + 6b \text{ for some integer } b. \qquad (15)$$

Step 3: Combining (13) and (15),

$$N = 5\ (4 + 6b) = 20 + 30b. \qquad (16)$$

Combining (16) with $N = 3$ (mod 7),

$$3 \text{ (mod 7)} = 20 + 30b,$$

and so

$$4 \text{ (mod 7)} = 30b. \qquad (17)$$

Step 4: To solve (17) for *b*, once again Euler's theorem is used. Here, the result is,

$$b = 4\ (30)^{\varphi(7)-1} \text{ (mod 7) where } \varphi(7) = 6,$$

$$b = 4\ (30)^5 \text{ (mod 7)} = 2 \text{ (mod 7)} = 2 + c \text{ for some integer c.} \ (18)$$

Step 5: Combining (16) and (18),

$$N = 20 + 30\ (2 + 7c) = 80 + 210c$$

$$N = 80 \text{ (mod 210)}.$$

The modern mathematical procedure makes the simplicity of the direct use of the tika all the more apparent. In large part, that is because the mathematical procedure can be applied more widely, while the tika is particular to these cycle lengths. But that is precisely the point—the tika is not a calculating device but is rather an elegant and parsimonious symbolic representation of the concurrent cycles in the Balinese calendar. By its spatial arrangement, and with a minimal set of symbols, it visually conveys all that is needed.

Another Balinese structuring of time involving the concurrence of cycles is found in the music of Bali, in particular the gong ensemble music that accompanies rites and ceremonies. The gong ensembles or *gamelan* orchestras are made up of gongs of different types and sizes and, hence, of gongs with different timbres and pitches. Particularly in archaic gamelan pieces still played today, or in more modern pieces in which there is an overlay of melodic elaboration, the underlying structure is intersecting and overlapping cycles. The instruments are struck periodically, each with a different period, with the times of striking so patterned that, together, they form a supracycle, which is then repeated.

Here, as an example, one such piece is described. We denote the largest gong by A and the two other types by B and C. D and E are both the same instrument but with higher and lower pitches. In linear order, the gongs struck are:

A&B, E, D, E, C, E, D, E,

– , E, D, E, C, E, D, E, B, E, D, E, C, E, D, E,

– , E, D, E, C, E, D, E, A&B, ...

Gong A marks the starting point of a supracycle. During the time of one supracycle, B is struck twice, C is struck four times, D is struck eight times, and E is struck sixteen times. The strikings of each gong are begun at such times in the supracycle that, with the exception of A and B whose coincidence marks the start of a new supracycle, none of the strikings are simultaneous. Figure 3.9 illustrates the same piece, but highlights where each gong is struck in the supracycle.

The supracycle is effectively subdivided into 32 parts. When looking at Figure 3.9, it is important to bear in mind that the sound made when a gong is struck persists for some time; longer sounding instruments are those struck less often in the supracycle. As a result, although the striking cycles within it are offset from each other, the sounds, nevertheless, do overlap.

While those of us accustomed to a different type of music might neither appreciate nor hear these cycles within cycles or cycles upon cycles, they are an auditory manifestation of the Balinese structure of time. In a similar way, the patterned symbols on the tika are a visual manifestation of the same structure. And the Javanese–Balinese calendar itself is a highly formalized and abstract conceptualization of the same temporal structure used to organize the activities of individuals and groups. All of these forms reiterate the Balinese concern for knowing clearly and precisely where an instant is situated in time

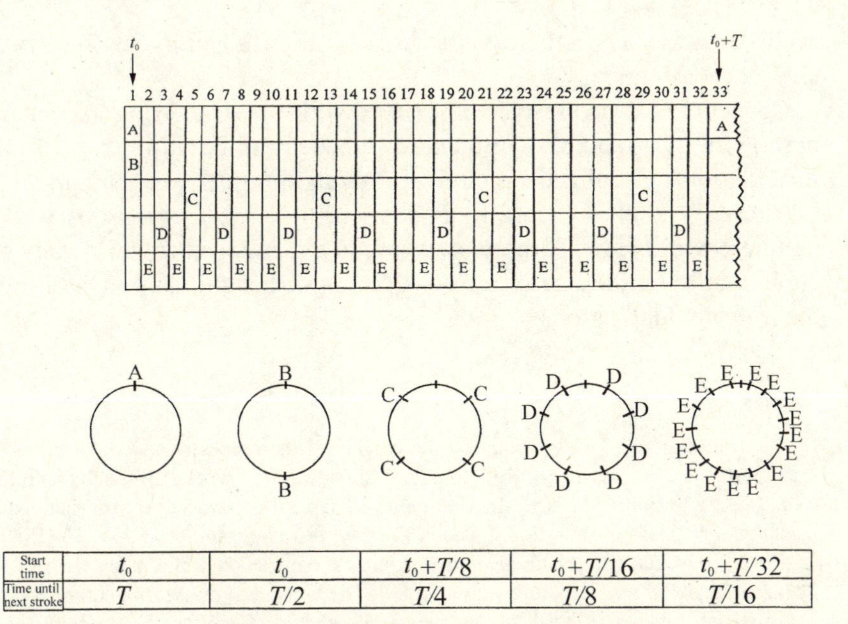

Start time	t_0	t_0	$t_0+T/8$	$t_0+T/16$	$t_0+T/32$
Time until next stroke	T	$T/2$	$T/4$	$T/8$	$T/16$

Figure 3.9 The chart on top shows linearly when each gong is to be struck. Below, the same information is shown in circular form. t_0 is the starting time for the piece, and T is the time for one full supracycle.

and, hence, where it is situated in the confluence of the controlling forces of the gods and demons of the Upper and Lower Worlds. Time is not simply to be noted: it is an active, rather than a passive, part of an event.

5 The Javanese–Balinese calendar is, perhaps, the most abstract of calendars. For those of us not enculturated as Balinese, even visualizing the interplay of ten concurrent cycles is not easy. Using them to regulate our daily lives would be even more difficult. There are calendars of numerous other cultures involving multiple cycles, but our discussion of the Javanese–Balinese calendar and Maya calendar should be sufficient to accentuate the fact that the purpose of a calendar need not be to remain in synchronization with the sun, or with the moon, or with any other physical cycle. The more divorced the calendar is from physical cycles, the more it becomes a creative expression of abstract mathematical ideas. For the Maya and Balinese, as well as for others whose calendars involve the interplay of abstract cycles, some of the mathematical ideas and questions are restricted to

specialists, but, nevertheless, the logic of interlinked cycles is pervasive in the cultures.

Concepts of time are part of a framework for structuring and interpreting the surrounding world and events occurring within it. As such, formulations, visualizations, and representations of time are an integral part of daily thinking. In all, concepts of time are a crucial part of a culture's world view. Although we cannot experience the world view of another culture, their calendar provides us with significant insight into their conceptualizations.

Notes

1. The spread of the 7-day week and the origins of the names assigned to the days within it are discussed in "Religions and the seven-day week," Boris Rosenfeld, *LLULL*, 17 (1994) 141–156. In "Naming the days of the week: A cross-language study of lexical acculturation," Cecil H. Brown, *Current Anthropology*, 30 (1989) 536–550, 148 languages are examined to show the diffusion of the 7-day week. The pair of articles give a picture rich in detail. The week as a cultural phenomenon is discussed in an excellent book by Eviatar Zerubavel, *The Seven Day Circle: The History and Meaning of the Week*, Free Press, New York, 1985. The chapters that discuss attempts to modify the week by the post-revolutionary governments of France and Russia give particularly interesting insights into the cultural hold of the week.

For discussions of the Akan calendar, see Philip F.W. Bartle, "Forty days: The Akan calendar," *Africa*, 48 (1978) 80–84; "The origins of the Akan," Abu Boahen, *Ghana Notes and Queries*, No. 9, Nov. 1966, pp. 3–10 and pp. 113–115 in *Ashanti*, R.S. Rattray, Negro Universities Press, New York, 1969 (a reprint of a 1923 publication by Clarendon Press). "The Northern Thai calendar and its uses," Richard Davis, *Anthropos*, 71 (1976) 3–32 is an excellent, detailed exposition of a calendar that has both cyclic and linear components and how the calendar is integrated into the life of the Muang people. An appendix contains 24 formulas for deciding whether days are auspicious or inauspicious for particular activities.

2. For excellent comprehensive overviews of the Maya, see J. Eric S. Thompson, *The Rise and Fall of Maya Civilization*, University of Oklahoma Press, Norman, 1954 and John S. Henderson, *The World of the Ancient Maya*, Cornell University Press, Ithaca, NY, 1981. Maya concepts of time and the role and significance of the calendar in daily life are discussed in "Remembering the future, anticipating the past: History, time and cosmology among the Maya of Yucatan," Nancy M. Farris, pp. 107–138 in *Time: Histories and Ethnologies*, D.O. Huges and T.R. Trautmann eds, University of Michigan Press, Ann Arbor, MI, 1995; *Time and Reality in the Thought of the Maya;* Miguel León-Portilla, Beacon Press, Boston, MA, 1973; and *Time and the Highland Maya*, Barbara Tedlock, University of New Mexico Press, Albuquerque, NM, 1982. A brief well-stated presentation combining both concepts and time measurement specifics is "Time in Maya culture," Henry J. Rutz, pp. 981–983 in *Encyclopedia of the History of Science, Technology, and Medicine in Non-*

Western Cultures, H. Selin, ed., Kluwer Academic, Dordrecht, 1997. Specific extensive, detailed discussions of Maya calculations are in "Maya numeration, computation, and calendrical astronomy," Floyd G. Lounsbury, pp. 759–818 in *Dictionary of Scientific Biography,* vol. 15, Supplement 1, 1978 and in "Mathematical notation of the ancient Maya," Michael P. Closs, pp. 291–369 in *Native American Mathematics*, Michael P. Closs, ed., University of Texas Press, Austin, TX, 1986. Also "The ancient Maya: Mathematics and mathematicians," by Closs in *Proceedings of the CSHPM 18th Annual Meeting 1992,* vol. 5, J.J. Tattersall, ed., 1992, pp. 1–13 discusses the archeological evidence regarding the scribes and their subgroups that worked with the mathematical calculations. The evidence shows that these groups included both men and women.

The Long Count date 9.0.19.2.4, which we used as an example, was on a stela that was also dated in the Calendar Round as 2D4, 2A10. According to one commonly accepted correlation with the Gregorian calendar, the beginning of the Great Cycle that was then ongoing was in 3114 BCE. Hence, the date given by the Long Count would be in 454 CE. The stela also placed the day within a 9-day cycle of *Lords of the Night*. (Each day in this cycle was associated with one of the nine levels of the underworld.) In addition, the day was placed within a lunar cycle. Lunar years and half-years are made up of 29- and 30-day lunar months. The stela contained the moon number within the lunar half-year, the age of the moon, and whether it was a 29- or 30-day month. On some stelae, the dates also identify a day within the 819-day cycle associated with the rain god.

3. For basic discussions of the algebra of congruences, see *Invitation to Number Theory*, Oystein Ore, New Mathematical Library, MAA reprint of 1967 original; *Elements of Number Theory*, I.A. Barnett, Prindle, Weber, and Schmidt, Boston, MA, 1969; and pp. 281–321 in *Invitation to Mathematical Structures and Proofs*, Larry J. Gerstein, Springer, Sudbury, MA, 1996. Some history and additional references for the Chinese Remainder Problem are in *A History of Mathematics: An Introduction*, Victor J. Katz, HarperCollins College Publishers, New York, 1993.

4. For excellent overview of Balinese culture, see the "Introduction," pp. 1–76, by J.L. Swellengrebel in the book he edited *Bali—Studies in Life, Thought, and Ritual*, W. Van Hoeve, The Hague, 1960; *Traditional Balinese Culture*, Jane Belo, ed., Columbia University Press, New York, 1970; *The Three Worlds of Bali*, J. Stephen Lansing, Praeger, New York, 1983; and Lansing's, *The Balinese*, Harcourt Brace, New York, 1995.

An interesting sequence of articles discussing the Balinese concept of time is *Person, Time, and Conduct in Bali: An Essay in Cultural Analysis*, Clifford Geertz' Cultural Report Series, No. 14, Yale University, New Haven, 1966 (reprinted in Geertz', *The Interpretation of Cultures*, Basic Books, New York, 1973); "The past and the present in the present," Maurice Block, *Man*, 12 (1977) 278–298; and "The social determination of knowledge: Maurice Block and Balinese time," Leopold E.A. Howe, *Man*, 16 (1981) 220–234. (The latter article contains the observation cited in the text about the Balinese solution of such problems as, "How many days to Galungan?").

Additional specifics about the calendar are in "Holidays and holy days," R. Goris, pp. 114–129 in the book edited by Swellengrebel cited above; Appendix I, pp. 223–239 in *Modern Javanese Historical Tradition*, M.C. Ricklefs, School of

Oriental and African Studies, University of London, London, 1978. The illustration of the paper tika follows the figure on p. 285 of *Island of Bali*, Miguel Covarrubias, Alfred A. Knopf, New York, 1938, and the wooden tika is described in detail in "Een Balineesche kalender," W.O.J. Nieuwenkamp, *Bijdragen tot de Taal-, Land-En Volkenkunde van Nederlandsch–Indie*, 69 (1914) 112–126. Figure 3.5 is from Plate I on p. 117 of this article.

Discussions relating Balinese concepts of time with music are "Time and tune in Java," Judith Becker, pp. 197–210 in *The Imagination of Reality*, A.L. Becker and A.A. Yengoyan, eds., Ablex Publishing Co., Norwood, NJ, 1979; "A concept of time in a music of Southeast Asia," José Maceda, *Ethnomusicology*, 30 (1986) 11–53; and "Epistemology and music: A Javanese example," Stanley B. Hoffman, *Ethnomusicology*, 22 (1978) 69–88. The specific gamelon piece described in the chapter follows his discussion.

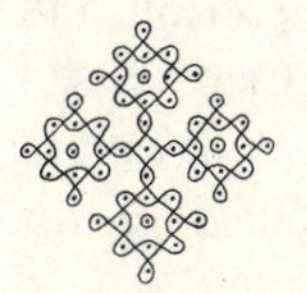

Models and Maps

The Marshall Islands stick charts first came to the attention of Westerners in an 1862 report by an American missionary. In his brief paragraph about them, he says that they were used to retain and impart navigational knowledge, but were so secret that his informant, although the husband of a chief, was threatened with death. During the next 50 years, about 70 charts and some information about them were obtained from Marshall Island navigators or those who claimed to understand these navigational aids.

Here, we examine these charts and the knowledge they embody in order to increase our understanding of the scientific and mathematical ideas of the Marshall Islanders. As so often when we look at ideas in cultures other than our own, examination of the ideas of the Marshall Islanders leads us to think further about some of our most basic concepts. In this case, we focus in on ideas about models and ideas about maps.

1

A standard dictionary definition of a map is "a representation, usually on a flat surface, of a portion of space." Generally, its purpose is to locate specific places in relationship to other places. Numerous conventions have been developed to give meaning to the word "locate" and to make a map more than a personal mnemonic device. A world map in an atlas, for example, includes longitudes and latitudes developed to provide a grid system that can be used to specify any particular point—that is, we locate a point at the intersection of two lines. Most maps also include for orientation the directions North, South, East, and West, and usually, distances between points are much smaller than, but proportional to, distances in the space being depicted. These conventions, however, are not the only ones possible. In fact, because the earth is essentially spherical, when a sufficiently large portion of it is depicted on a flat surface, both direction and distance cannot be preserved. For example, the familiar maps based on Mercator projections are conformal maps, meaning that angles are preserved. But, as a result, shape and distance are distorted. Even for small regions, quite different conventions can be used.

In our daily lives, we encounter a variety of maps. Most of them are used to enable us to get from one place to another. Others, such as weather maps or geological maps, serve to convey data visually in a meaningful way. The latter correlate different types of information with spatial locations. The meaning of the data, whether it be temperature or air pressure, soil and rock type, or demographic statistics, and the symbols used to convey it, must be specifically learned for the map to be understood. There is no clear demarcation between maps for travel and those for data display. In fact, many travel maps contain diverse information about the regions through which one may pass.

Let us look more closely at travel maps which, because they have a specific function, can be discussed in relation to their function. Travel maps, first of all, vary considerably, depending on the means of travel for which they are intended. Maps for automobile travel, air travel, hiking, subway travel, ocean voyaging, or coastal sailing, for example, are quite distinct from each other. Not only do these maps differ in the types of information they contain, but they differ in level of detail and the size of the region depicted. Their specific contents are selected to suit the means of travel. It would be inappropriate, if not impossible, to use a subway map for air travel or a road map for sailing.

Figure 4.1 shows a portion of a map used for hiking in the vicinity of Acadia National Park on Mt. Desert Island in the state of Maine in the U.S.A. In addition to indicating landmasses, park boundaries, roads, bodies of water, and streams, the map shows the location of hiking trails, as well as mountain summits and their elevations. Also, an arrow on the map shows geographic north, and there is a scale showing the length that represents a mile. Most distinctive, however, the map has contour lines—that is, a set of closed curves in which each curve connects locations that are the same height above sea level. The set of curves are drawn to show intervals of 50 feet of height. These contour lines can also be thought of as where a set of planes, each separated from the one below it by 50 feet—all parallel to a hypothetical plane at sea level—would intersect the irregular, three-dimen-

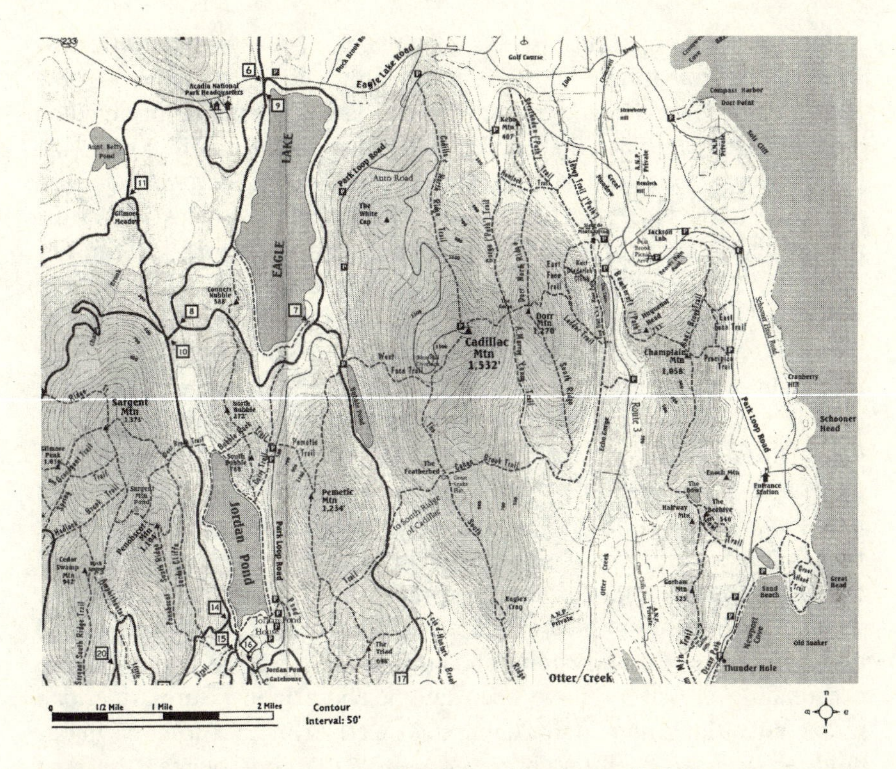

Figure 4.1 Hiking trail map: Acadia National Park. (© 1998 Tom St. Germain, Parkman Publications. Reproduced with permission.)

sional landscape. Thus, where lines are close together, the land rises steeply; and where lines are far apart, there is only a very gradual change in elevation.

In all, to utilize this map, hikers must be familiar with the kinds of land and water features included, but also must understand the meaning and use of a scale, the *feet, mile*, *sea level* system of measurement, and the meaning and implication of contour lines.

In contrast to hiking maps, maps intended for automobile travel generally cover a much larger region. Quite often, the region is politically defined, such as, within the U.S.A., a state or a group of states, or a county within a state. These maps, too, indicate landmasses and bodies of water, as well as political boundaries within the region. In addition, they mark national, state, and local roads; population centers; controlled access roads and their interchanges; multi-lane and single lane roads; paved and unpaved roads, and perhaps other features of importance to auto travelers, such as rest areas. Distances and a scale of miles are very important aspects of the map. Most include an arrow somewhere indicating geographic north, although, quite frequently, the need for this is superseded by the convention that the map is oriented on a rectangular page such that up is north, down is south, right is east, and left is west. Longitude and latitude are rarely, if ever, included on highway maps. However, in order to locate specific places within the region, the mapmaker usually superimposes his own grid system. For example, by placing equally spaced numbers along the top and bottom edges and letters along the sides, any specific place in the region can be identified as in the rectangle centered at, say, the point B7; that is, in the rectangle centered at the point where a line perpendicular to a side edge at B intersects the line perpendicular to the top or bottom edge at 7. Quite often, maps of smaller regions are inset in, or in some way appended to, a road map. These smaller regions may be denser in roads or of special interest to travelers, thus requiring a larger scale. In general, there is no one size or scale that is appropriate for all travelers and all regions, and so a number of maps are often used. Despite their differences, we can read them because of our knowledge and experience of road systems and auto travel. Even more important to our reading is that we have been taught about many Euro-American map conventions, and, in this context, as well as in other contexts in our culture, we have been taught about scales, the units of measure, and grid systems, and also about political subsubdivisions and jurisdictions.

Before moving to a general definition of maps, let us look at one more type of travel map from our culture; this type is quite different from the hiking maps and auto travel maps. A striking difference is that this type of map has no precise scale and no distance indicators. In fact, it has little else that we might commonly expect on maps. Figure 4.2 is a map of the metro system in Washington, DC (U.S.A.). Such maps are familiar to those who live in or visit these cities or other densely populated areas including New York City, Mexico City, Munich, Paris, London, or Tokyo.

Regardless of the cities they come from, metro maps are remarkably similar in content and in their form of representation. They show, essentially, a set of named points interconnected by lines of different colors. The use of color is particularly important as it indicates which of the line segments connect to each other to form a continuous path linking a subset of the points. Generally, two or three different types of points are distinguished from each other in some way. The points, as we have learned, represent train stations, and the lines represent the paths of the

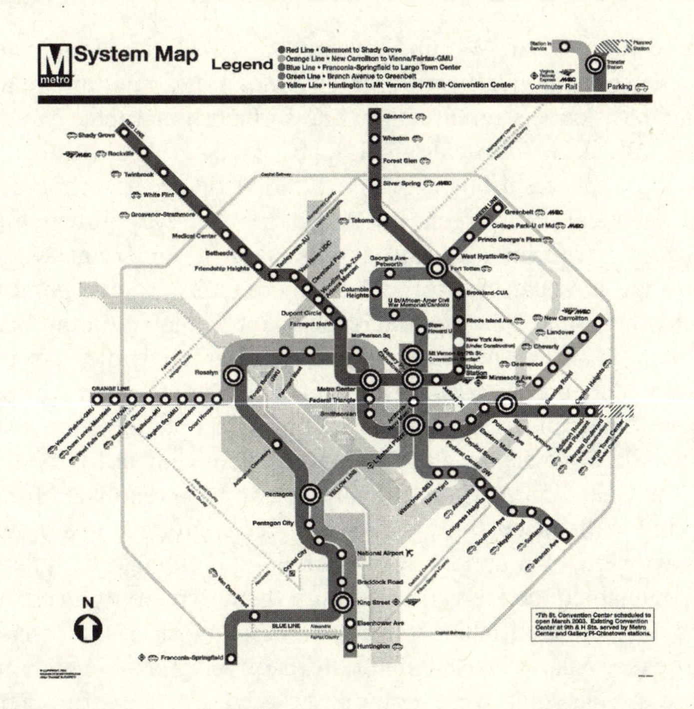

Figure 4.2 Washington, DC Metro system map. Figure courtesy of the Washington Metropolitan Area Transit Authority.

trains. Each color is a different train line, with the points along it being the stations at which that train stops. The specially marked points are stations where a passenger can transfer from one train line to another. Although not indicated on the maps, it is understood, unless otherwise noted, that there are trains that go in opposite directions along each path. Because the maps represent particular configurations in space, to some degree, the layouts maintain the spatial relationships of the physical parts of the metro networks. While quite imprecise with respect to distance or directions, metro maps contain all that one needs to know to get from here to there within the metro network. To a metro rider, which train line a station is on, or at which station one can change from one line to another, is far more significant than the shape of the landmasses or the exact distances being traversed.

Maps, as just these few examples show, must be viewed broadly; and, with a broad view, we can better appreciate them as products of mathematical abstraction. The mapmaker draws upon diverse information that has been obtained from drawings, reports, and stories; and from these, he creates an analogical space—that is, a space that is substituted for another. Within the space, physical entities selected as significant are symbolically represented. And, most important, relationships that may not be seen directly while *in* the original space, are made visually explicit in the analogical space. It is in establishing these relationships that scientific or technical knowledge is used (or created), and mathematical ideas are an integral part of the way in which these relationships are formulated and expressed. It is important to realize that maps are not snapshots of space—there is no vantage point from which anyone can *see* the original space as it is shown on the map.

Although, in Euro-American culture, maps are generally associated with their paper embodiments, maps can be rendered using other media as well. In other times and in other cultures, people have used stone, clay, treebark, or whatever was available and convenient for them. The Inuit of the Canadian Arctic, for example, who are renowned for their mapmaking skills, used snow- or sand-covered ground. The drawings themselves were soon gone, but their creation combined with detailed discussion enabled the viewers to commit them to memory. And, as we shall see, the Marshall Islanders, who live just north of the equator in the Pacific (see Map 4.1) used materials readily available in their environment–materials that were commonly used by them for creating durable artifacts.

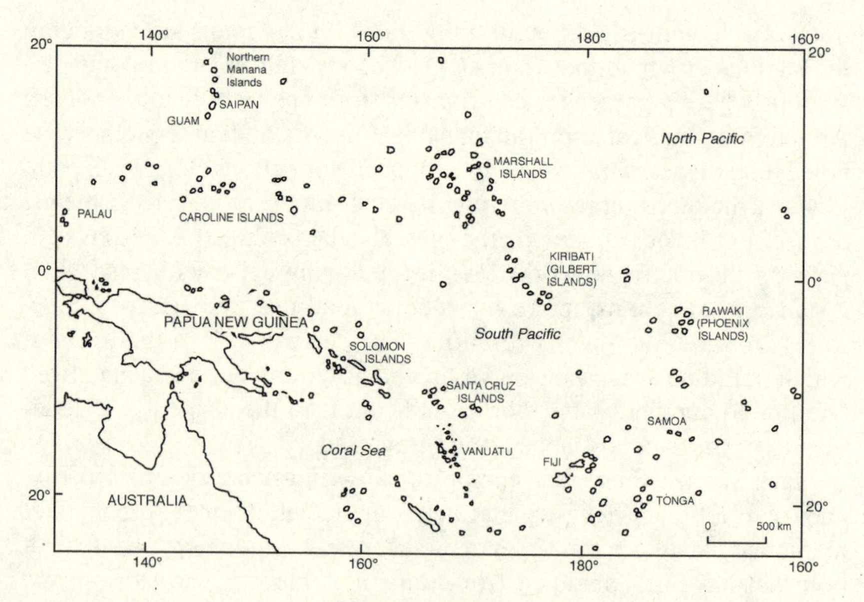

Map 4.1 Oceania.

In the history of Western cartography, a distinction was made between maps and charts. Charts referred to the depictions used by mariners that contained varied types of information based on their experience and specific to their purposes. Maps, however, were largely academic, concerned with the world as a whole. Early cartographers, such as Ptolemy of Alexandria, Greece (ca. 120 CE), defined what they did as *geography*–"a representation in pictures of the whole known world together with the phenomena which are contained therein." He distinguished that from *chorography*, which he deemed regional and selective, "even dealing with the smallest conceivable localities, such as harbors, farms, villages, river courses, and the like." Our broader definition of maps is in keeping with more modern writers who view world-wide maps and local maps simply as different streams, which have an underlying conceptual unity and which eventually merged. Differences in terminology, however, have persisted. Hence, maps specifically for mariners are still called charts, and so the unique objects created by the Marshall Islanders are commonly referred to as *stick charts*.

Made of palm ribs tied with coconut fibers and sometimes with a few shells attached, the Marshall Islands stick charts are sizeable objects, generally about 60–120 cm by 60–120 cm. The stick charts are, essentially, of two different types. One type is maps. Figure 4.3 shows what a few of those that are maps look like. For a long time, Westerners did not recognize these as maps and found them quite perplexing. The Inuit maps, by contrast, although from another culture, were easily recognized and considered to be very accurate and perceptive. As our maps did, the Inuit maps contained details of the coastline and distinguished waterways from interspersed landmasses. Despite the fact that the Inuit means of travel and modes of navigation may have been considerably different from that of the Westerners, the drawings that the Inuit made for them were sufficiently similar to be understood. The Marshall Islands stick charts, however, are renderings of large open expanses of ocean. What they show is meaningful *only* if one knows something about *wave piloting*, a system of navigation unique to the Marshall Islands. As contrasted to, say, our road systems or metro systems, which rely on material objects fixed in space, the wave piloting system depends on a conceptualization of the dynamic interplay of land, wind, and water. We have learned, for example, to distinguish between multi-lane controlled access highways and unpaved single lane roads, and how they affect automobile travel. But whether or not we use the roads

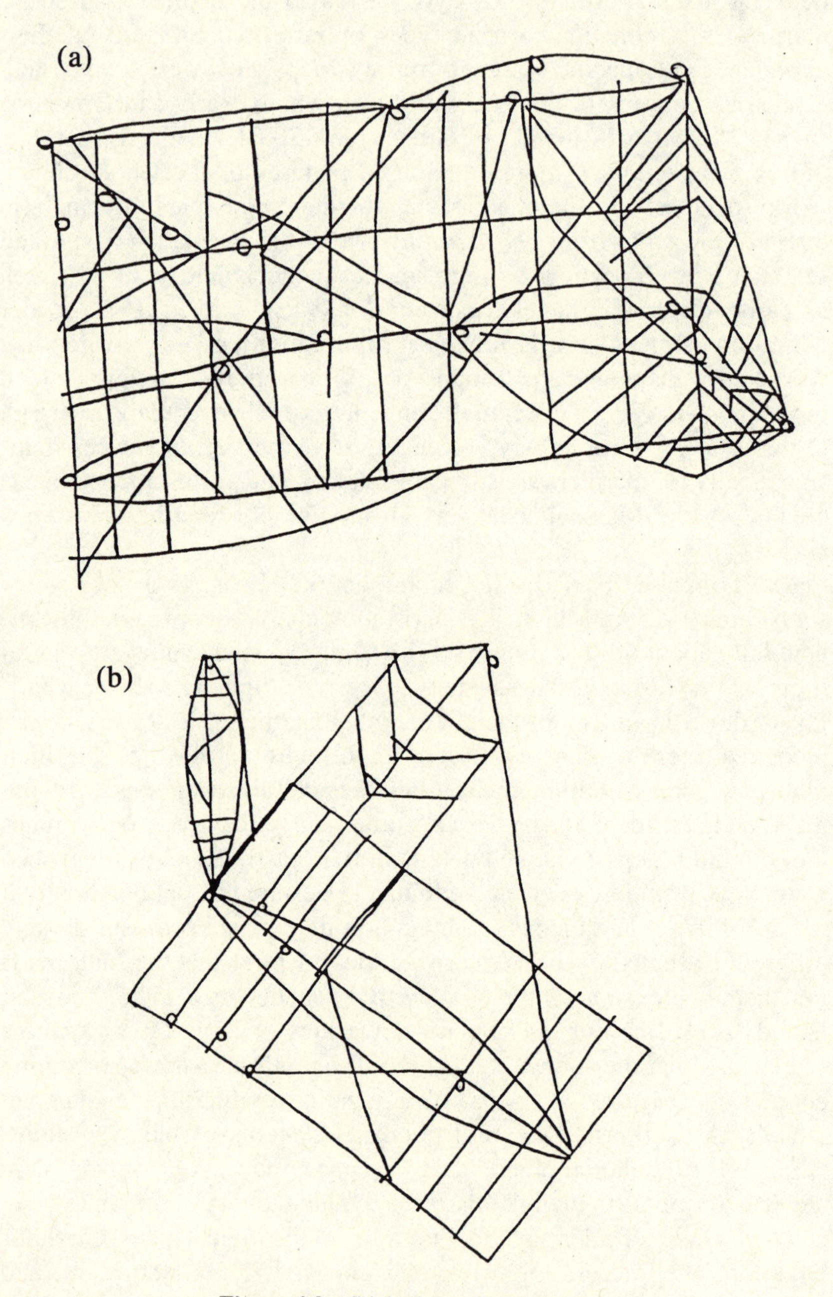

Figure 4.3 Stick charts that are maps.

for automobile travel, they and their distinctions, nevertheless, are there. Similarly, to Marshall Island navigators, the features depicted on their maps, although they are *not* concrete objects and *not* fixed in space, are present in the ocean, whether or not one has learned to see them or use them.

The stick charts that the Marshall Islander's refer to as *meddo* and *rebbelith* are maps. The other stick charts, the second type, are called *mattang* and are not all what we classify as maps. Used as training devices, the mattang introduce the prospective Marshall Islands navigators to the features of the environment that will be included on the maps. They show the interplay of oceanographic phenomena and land masses. That is, they are static, idealized representations of shapes and motions in the sea and at the land/sea interface. The mattang are explanatory models of the dynamic geometry that underpins wave piloting.

Thus, the mattang are the key to gaining insight into the system that utilizes the stick charts that are maps. As is the necessary prerequisite for reading any map, we (and the Marshall Islands navigators) must first understand what in the original space is considered significant and so will be preserved in the analogical space. And of course, as one learns what is significant, one is learning why it is significant or what role it plays in the system. But over and above being necessary for map reading, the mattang are of importance in themselves, because of the mathematical and scientific ideas they embody, and because they are exemplars of explanatory models. Before we go further, however, we need a greater understanding of the surroundings in which the stick charts were created and used.

2 The Marshall Archipelago consists of about 29 coral atolls and 5 small coral islands formed into two parallel chains running about 960 km in a northwest–southeast direction (see Map 4.2). Each atoll consists of a lagoon surrounded by a narrow ring of coral reef and small islands. Lagoons of larger atolls are in the range of 32–42 km in length and 8–16 km in breadth. Bikini, the northernmost atoll in one of the chains, became famous as the site of U.S. nuclear bomb tests after World War II. It consists of a ring of 51 small islands with a total land area of 7.7 sq. km surrounding a 630 sq. km lagoon. Before they were moved in 1946, the Bikini population was 170 people. Another example, in the southern end of the other chain, is the more densely populated Majuro atoll. Consisting of a ring of 61 small islands surrounding a 260 sq. km lagoon, its total land area is about 8.5 sq. km, and in 1946, it had a

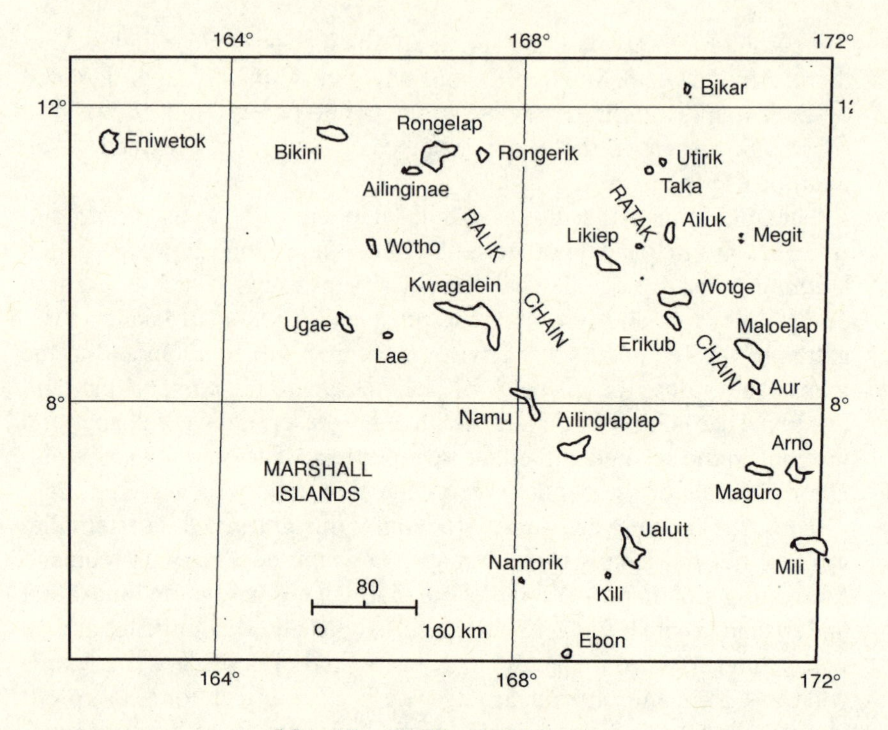

Map 4.2 The Marshall Islands.

population of about 1600 people. In all, the Marshall Islands have a land area of just under 180 sq. km scattered over about 970,000 sq. km of ocean. At the time they became part of the U.S. Trust Territory in 1947, there was a population of about 10,000 people.

Because there is so little land, with much of what there is being poor soil along the shores or swamps in the interiors, considerable community attention focuses on how land is passed on, how it is used, and how its fruits are distributed. Although there are differences in details and in formality from atoll to atoll, there are clear social classes. Within these classes, there are further distinctions based on maternal lineages (individuals who are related through successive mothers to a common ancestor) and on clans (groups of related lineages). On Majuro, for example, about 10% of the people are nobility, while the other 90% are commoners. Land is never sold and is never personally owned by a single individual. Although it is worked by the commoners, rights to the produce of any piece of land are shared by the paramount chief, a noble lineage, and a commoner lineage. When possible, small surpluses are given to those who provide specialized services, such as shamans, navigators, and master craftspeople. The shore areas grow coconut palms, and pandanu; further inland grow breadfruit, arrowroot, banana trees and some hardwoods; and taro grows in the swampy interior. These provide food and fuel, as well as construction materials and weaving materials. Sailing canoes, for example, were made from logs of the breadfruit trees lashed together with lines spun from coconut husk fibers. The woven objects, made primarily from pandanu leaves and coconut leaves, included sleeping mats and highly decorated women's dress mats, as well as sails for the sailing canoes. And copra (dried coconut meat) has been an export since the coming in 1860 of the Americans, then the Germans, then the Japanese.

It is water, however, that dominates the environment. Hence, sailing and boats are an integral and essential part of life. On an atoll, within the lagoon, sailing canoes are used for fishing, for the collection of food and copra, and for traveling to visit friends. In addition, boats are used for open sea fishing in the vicinity of the atoll and, beyond that, for open sea travel to other atolls. Some small sailing canoes of about 5 m, which can also be paddled or rowed, are for use around the edges of the reefs. But to cross the lagoon, and to carry passengers and cargo, the boats are 7–9 m long and require a crew of two or three people. These boats, and still larger ones, are used for travel to other atolls in the Marshall archipelago and for trips that extend even further.

The boats used by the Marshall Islanders are generally referred to as *outriggers* because of their distinctive configurations. As contrasted to the sailing boats more common in Euro-American environments, the Marshall Islands boats are asymmetrical with respect to the center line from bow to stern. One side of the hull is flat, and the opposite side is convex. Projecting outward from the convex side is an outrigger made of poles attached to a shaped log. On the poles, adjacent to the hull, there is a platform on which the crew sits (see Figure 4.4). In larger boats, there is a small hut on the platform providing an enclosed space for provisions, sleeping mats, and travelers. As a result of the boat's asymmetry, in sharp contrast to Euro-American sailboats, the sail can be put out only on one side, that is, on the side opposite the outrigger. Further, on Euro-American sailboats, there is a distinct difference between the front end and the back end; generally, the bow is wedge-shaped, and the stern is flattened. The Marshall Islands outriggers, however, are canoe-shaped; that is, both ends are wedge-shaped so that either end can serve as the front or as the back. These differences in design, of course, lead to differences in how the boat is sailed.

Figure 4.4 An outrigger on Jaluit Atoll. (Photographed in 1899 by H.C. Fassett. National Archives photo no. 22-ss-104b.)

Still being taken in the first part of the twentieth century were long-distance sailing trips to such places as the Caroline Islands, and the yet more distant Palaus and Saipan. Both of the latter are over 2500 km from the Marshall Islands. Earlier, as well as long-distance trips by a few boats, there were large-scale sailing expeditions with numerous boats and hundreds of people. The 1862 report of the American missionary describes boats that could carry 50–100 men in the open sea, and also recounts the festive return from some northern atolls to Ebon of 800 people in a fleet of 40 boats. Another nineteenth-century account tells of 18.3 metre-long boats carrying 40 to 50 people each and flotilla of up to 100 boats. These large sailing expeditions were led by chiefs, but their success, and the success of the more common trips within the atoll chains, depended on the navigators.

The navigators, usually relatives of a paramount chief, were highly regarded, specially selected, and extensively trained. Their knowledge and techniques were greatly prized and well-kept secrets. Knowledge, in general, was viewed as a personal possession, and, as such, it was not freely shared or given away. The ownership of knowledge, however, carried with it the responsibility for its preservation and transmission. Thus, the navigators each selected some individual to whom they passed on their knowledge and their personally developed systems. The person selected was a favorite child or someone specifically adopted because of showing special interest or special aptitude. In addition, some navigators, who were considered masters, oversaw the teaching of the prospective navigators for their extended families. This gave rise to shared systems and ongoing schools that traced back to a master navigator. As a result, while there was *a* Marshall Islands navigation tradition, there were, within it, some differences from school to school. The child selected by a navigator to carry on the knowledge could be a male or a female, and at least two masters cited by the Marshall Islanders were women. There was, for example, a large school on Namorik that was said to have grown up around the master Legemugidj. He adopted Lidérmelu, who learned all her seafaring skills from him, and from then on, the school was known by her name.

Eventually, some of this navigational knowledge was told to Westerners. However, what was told was far from complete as the Marshall Island tellers never intended to give full understanding to others. Since the Marshall Islanders had no indigenous writing system, we have only what was eventually recorded by outsiders. We also have about seventy stick charts that still remain in museums and in private collections.

3 The extensive training of the navigators centered on wave piloting. The navigators had to learn to see and to understand the shapes and motions in the sea that would be used to guide their travel. They had to learn to read the stick charts that were maps, but first, and even more crucial, they had to learn what of the environment was considered significant and what role it played in their navigation system. And it was, as we noted, the stick charts referred to as mattang that played an important role in this training of the navigators.

A typical mattang is shown in Figure 4.5. That these are not simply personal idiosyncratic devices but, instead, formalized and standardized models, can be seen by comparing them with other, almost identical mattang that were separately reported and collected (see Figure 4.6). Some other examples are shown in Figure 4.7. When looking at these, several features are striking. One, of course, is their symmetry. Another is the interplay of geometric forms bringing to mind familiar words such as triangles, sectors, arcs, perpendicular bisectors, angles of intersection, points of intersection, and so on. Also, there is their diagram-like clarity. To emphasize that this is, indeed, the nature of the artifact and not the result of our rendering, we note in particular that Figure 4.5 is a photograph of a mattang, while Figures 4.6 and 4.7 are drawings. The lettered labels, however, are our superimpositions.

The various parts of the mattang have been described differently by different writers who spoke with Marshall Islanders. From these descriptions, it is clear that the mattang are generalized configurations containing idealized shapes and forms that were used to explicate the

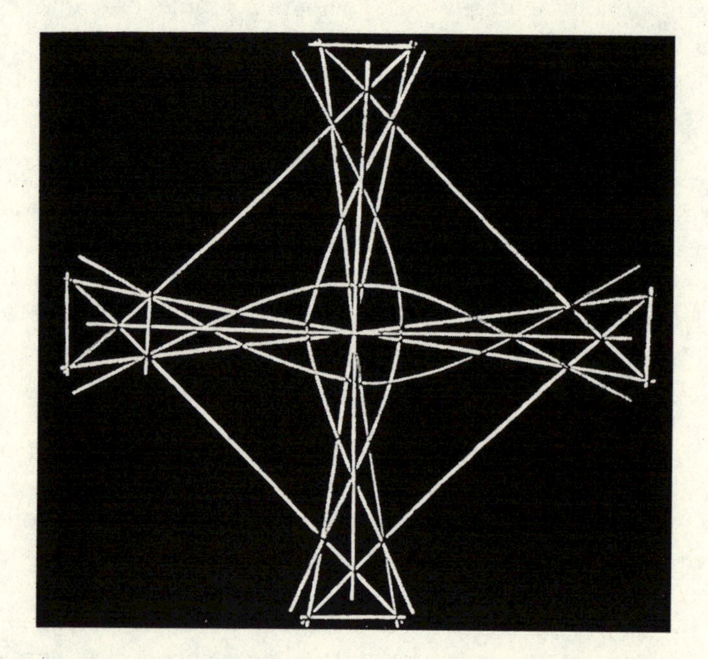

Figure 4.5 A typical mettang. (Plate VI in "On sea charts formerly used in the Marshall Islands, with notices on the navigation of these islanders in general," Captain Winkler, Annual Report of the Smithsonian Institution for the Year Ending June, 1899, Washington, DC)

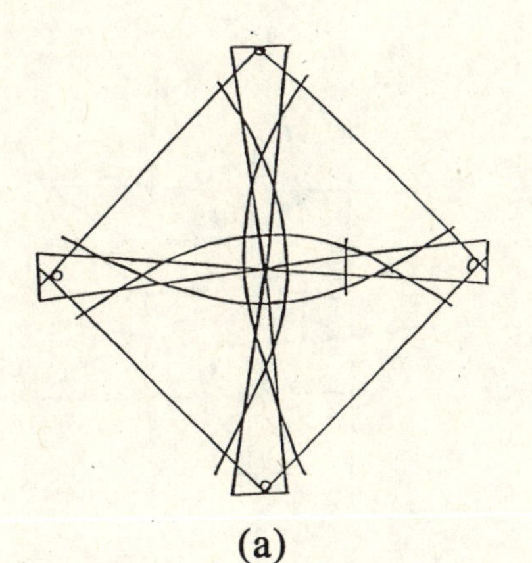

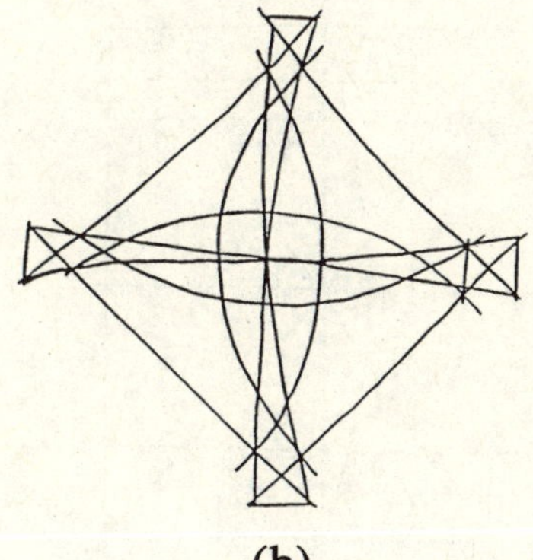

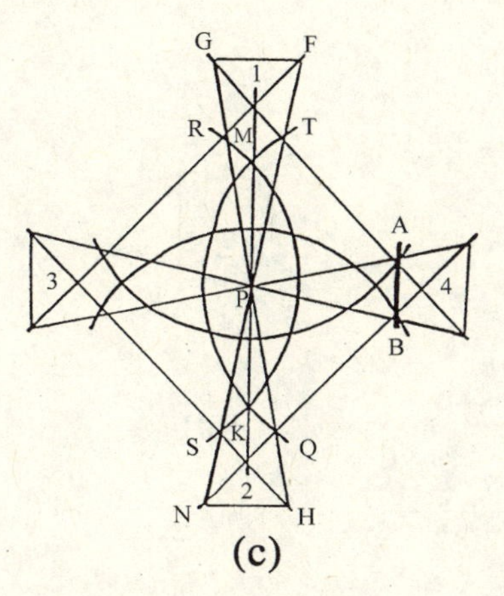

Figure 4.6 Similar mettang.

principles of swell and land interaction. What sometimes appeared to interviewers as confusion or inconsistency was, instead, a problem created by their own persistence in believing that the mattang were specific and concrete. In a telling exchange, we read, for example, that a Marshall Islander first associated some point with Jaluit atoll and later with Namorik atoll. When faced with this contradiction "the natives were quick to explain that it didn't matter—the chart was not where Jaluit in particular lay, but where land was. They stoutly maintained that it could be some island in another part of the world, one they had never seen." This, of course, is what we deem an idealized model. What is more, as a generalized illustrative configuration, parts of the mattang were referred to in different ways depending on the point

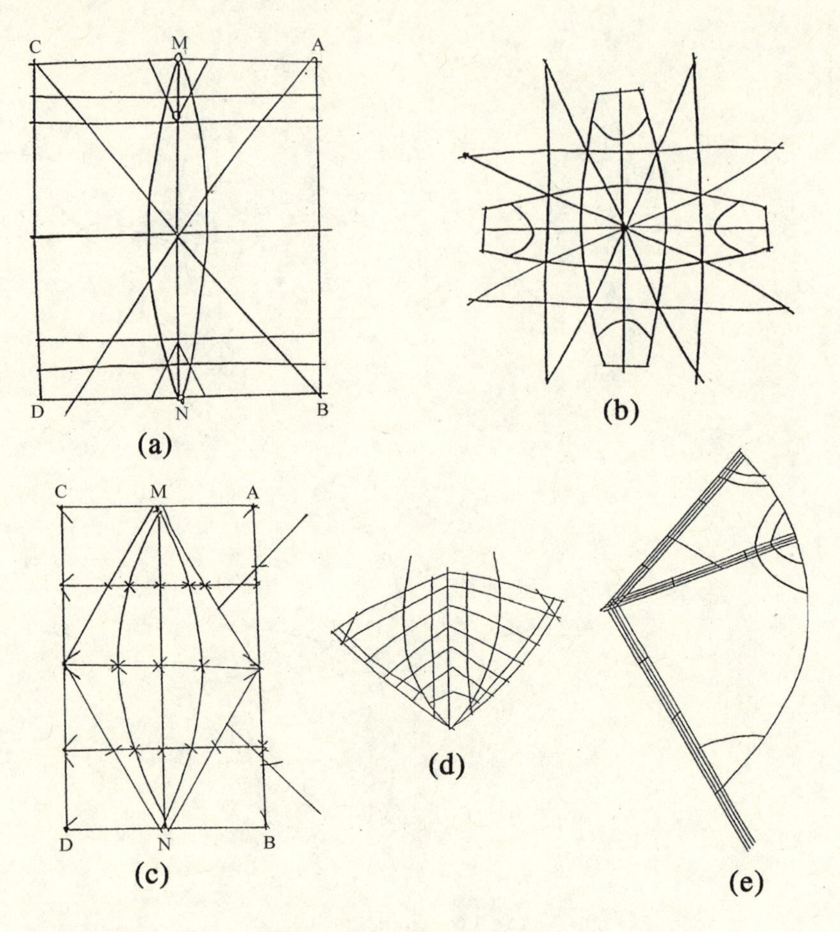

Figure 4.7 Other mettang.

being made. Rather than inconsistency, the various descriptions, taken together, convey a corpus of ideas.

Before looking at the mattang specifically, we must introduce some of the oceanographic phenomena fundamental to the system of wave piloting. Basic among these phenomena are what we term *refraction*, *reflection*, and *diffraction* of swells.

As waves move away from the winds that create them, they merge into swells formed by groups of waves of similar period and height. These sinusoidal swells can travel thousands of miles across the deep open ocean with very little loss of energy. For the open Pacific Ocean, the period of the waves—that is, the time interval between when one wave crest and the next pass some arbitrary point—is 16 or more seconds. Based on the period, the wave speeds can be calculated. For waves of period T, the wave speeds are $gT/2\pi$, where g is the gravitational constant 9.8 m/s^2. Hence, for these with $T = 16$ or more seconds, the wave speeds are more than 80 km/h. Swells move at about half the speed of the waves that they contain because some of the energy goes into setting the water in front of them into motion. Because the Marshall Islands are surrounded by deep (4000–5000 m) open ocean, long, fast-moving swells that are clear and consistent in pattern move toward them across the water. Swells, however, change when they meet underwater obstructions or reach shallow water. Shallow is a relative term, depending on the length of a wave: specifically, shallow is defined as less than half the wave length (wave length $= gT^2/2\pi =$ period $\times$ wave speed). In this case, that is upwards of 198 m. Hence, in and around the Marshall archipelago, the approaching swells are modified in direction and energy. The complicated and distinctive interactions of modified swells are the "landmarks" that the Marshallese navigators learn to read and interpret.

Wave refraction and its effects dominate the mattang. When waves move into shallow water, friction causes them to slow down. Depending on the ocean depth beneath it, a wave slows down differentially and so bends, eventually becoming more or less parallel to the underwater contours and, then, more or less parallel with the shoreline. This gives rise to the familiar observation of standing on any beach and seeing the waves come in toward you, even though further out they may be seen approaching at an angle. Figure 4.8 shows how a wave train wraps around a circular island assuming that the island has a uniform and gradually sloping underwater topography. Some energy is lost to friction, but concomitant with the bending of the wave front, its energy may become spread out or concentrated in different places, and there can be an increase in wave steepness until the wave peaks, becomes unstable, and breaks.

First of all, however, in the open sea, there is a swell moving in front of the wind. Let us say the wind is coming from the east. An atoll off in the westerly direction is signaled when the swell begins to bend inward. However, due west of the atoll, there would be a region where the refracted arms of the swell cross, although with greatly lessened energy.

In our refraction diagram (Figure 4.8), we showed a circular island with gradually sloping underwater topography. If the island were, instead, rectangular and rose up suddenly and steeply, the waves would be more abruptly stopped in their forward motion. But as the

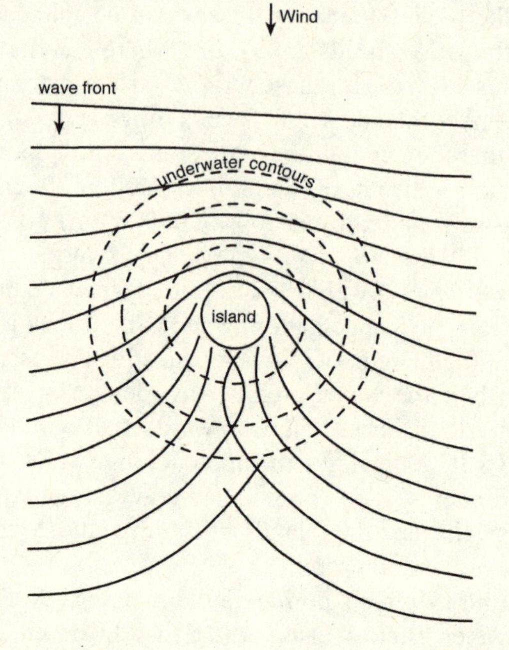

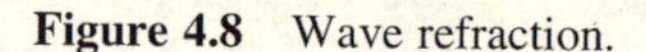

Figure 4.8 Wave refraction.

waves continue forward to the sides of the island, some of their energy would be propagated sidewards causing them to spread out along their edges. Just around the sides of the island, there may be an increase in wave height, but beyond that, these spread-out edges are increasingly lessened in height. As with the diffraction of light, there is a "shadow" behind the island, but it is a shadow with imperfect edges (see Figure 4.9). Thus, the more elongated the shape and abrupt the rise of the atoll, the less bending of the swell occurs on the windward side; on the lee side, the bent arms may not even cross as they are separated by a region with no energy from the swell.

Whether the land barrier rises gradually or steeply, some part of the wave will be reflected backwards. For a steep barrier, where there is no bending of the wave front, the reflected wave has much the same energy

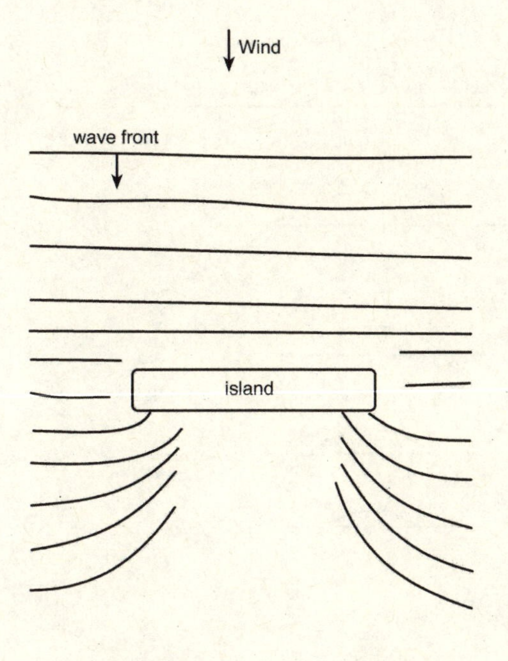

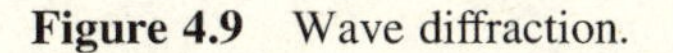

Figure 4.9 Wave diffraction.

as the incoming wave, and so, meeting quickly and head on, they spray water up into the air. For a gradually rising barrier, it is the bent wave that is reflected back to meet the incoming wave (see Figure 4.10). Hence, as waves meet at an angle, a running crest develops. This interaction of reflection and refraction takes place on the windward side of an atoll.

Let us now look at the mattang and their idealized geometry of swell interactions. On Figure 4.7e, one swell is represented; on Figures 4.7a and 4.7c, there are two opposing swells, and on the most common mattang (Figures 4.5 and 4.6), there are four swells, one from each of four perpendicular directions. Despite the symmetry of the mattang, generally there is a slight difference that specially marks one direction. On each mattang in Figures 4.5 and 4.6, there is a short stick (labeled AB on Figure 4.6c) wrapped in a cord showing the direction from which the prevailing wind is moving toward that central atoll (P). This serves as the orienting direction. Referred to as *rear*, the swell shown moving in front of this wind is *rilib* ("backbone"); the one

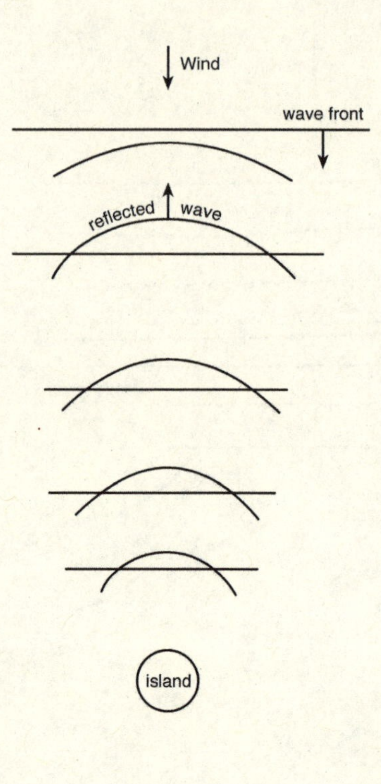

Figure 4.10 Wave reflection.

opposing it is *kaelib*; and the two from the perpendicular directions are *rolok* and *bundockerik*. Although *rear* is generally translated as "from the east," and the opposing and perpendicular directions then referred to as "west," "north," and "south," these words are simply for convenience. Their use does *not* imply that directions are conceived of in our frame of reference, and "east," "west," "north," and "south" are *not* the same directions as ours indicated by those names. I will follow this conventional translation as it simplifies referring to the direction the wind is coming from and the opposite and perpendicular directions. (In a few places, where it is significant, I will note the actual direction according to our system.)

In any discussion of sailing, be it ours or Marshallese, of paramount importance is the relationship of where a boat is, its destination, and where the wind is coming from. Western diagrams are generally circular and symmetric but, as with the mattang, they specially mark the direction of the wind. Thus, for both, that is the orienting marker to which all else is referred.

For Westerners, boat positions within the circle, and modes of sailing to a destination at its center, are discussed in terms of sectors (see Figure 4.11). Although the mattang are not circular, it is not surprising that they similarly have sectors prominently marked in some way. However, the differences between Western and Marshallese boats modify the angles of the sectors of interest. In contrast to a Euro-American boat, in which the sail can be put out on either side, as was noted in the previous section, on a Marshallese boat, the sail can only be put out on the side opposite the outrigger. Further, a Euro-American boat can head only one end, its front end, in the direction of travel, while for the Marshallese boat, either end can be its front end. *No* sailboat can sail directly into the wind (along the line from E to X on Figure 4.11). In fact, it cannot sail in a direction too close to that either. The boat would make little headway and might even be pushed back-

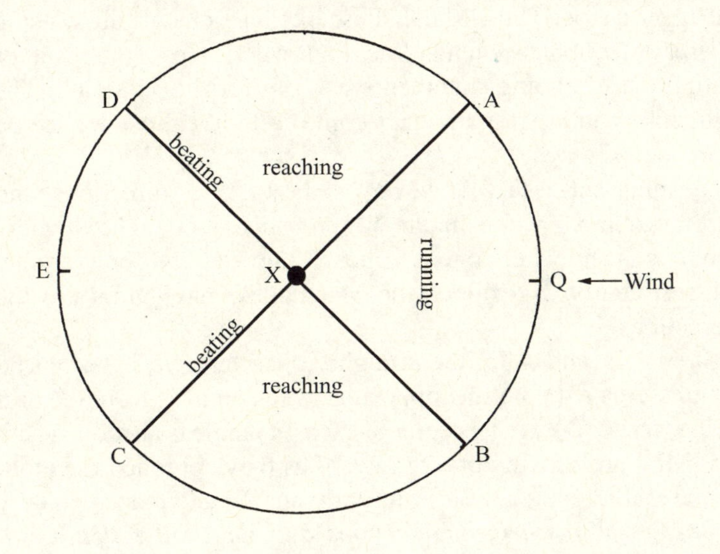

Figure 4.11 Sectors of sail.

wards. For Euro-American boats, *too* close is about 45° on either side of the wind (within sector DXC on Figure 4.11); for outriggers, it is within about 65°. In order to sail in those directions, it is necessary to *tack*, that is, to sail back and forth across the wind, *changing the side* on which the sail is out each time the direction is changed. To accomplish this on a Marshallese boat, the rigging and sail must be picked up and moved to the other end of the boat, and what was the front must become the back, and vice versa. Hence, they avoid, if possible, approaches from this sector. Furthermore, for them, the sector would be about 130° rather than our 90°. Similarly, in sector AXB, where the wind is behind the boat, the side on which the sail must be out differs for approaching X from within AXQ or from within QXB. Euro-American boats avoid a direction too close to QX since, if there is a slight deviation in wind direction or sailing direction, the full force of the wind can unexpectedly swing the sail across the hull to the other side. Sailors do, of course, cross the line and change the side of the sail when necessary, but, then, they do so with great care. For *either* type of boat, it is far preferable to remain decidedly on one side or the other of QX and so to have a course at least 15° or 20° away from it. Finally, we note that because of ease of steering and greater efficiency in a strong wind, a course perpendicular to the wind is said to be easiest and most efficient for an outrigger canoe. Thus, Marshallese explanations of the mattang place the greatest emphasis on that line. Figure 4.12 shows the result of incorporating the foregoing statements into our form of diagram. The marked directions and sector angles become quite like those we see on the mattang (see Figures 4.5–4.7).

Another important issue, however, is how these directions and sectors are correlated, when in the water and out of sight of land, with where one is and where one is going. While we rely, for example, on compasses, charts, and rulers, the Marshallese navigators use the swell interactions.

On Figures 4.7a and 4.7c, the straight edges that we have placed vertically on the page depict the opposing swells coming from the east (AB) and the west (CD). At the center of each rectangle is an atoll. Then, the swells (MN) are shown curving inward as they approach the atoll. Since, on the mattang, the arcs are similarly curved and placed equally distant from the atoll, *their intersections and the atoll fall along a straight line perpendicular to the direction of travel of the swells.* These intersections, and the north–south line that they define, are, as we noted, of paramount importance. (These lines and curved swells are

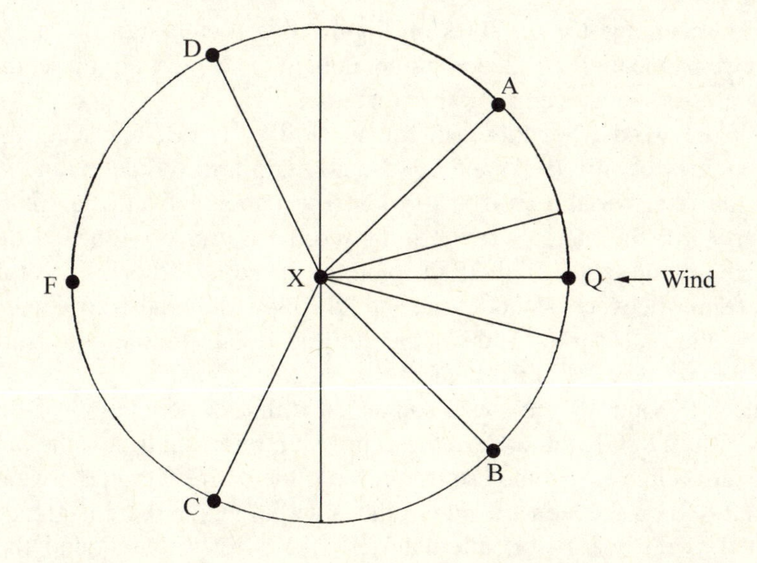

Figure 4.12 Significant directions and sectors when sailing toward X.

shown in a slightly different way on Figures 4.5, 4.6c, and 4.7b.) Where the bent arms of opposing swells cross, there is a narrow sector (most clearly seen on Figures 4.5, 4.6a, 4.6b, and 4.6c) in which a series of wavelets can be observed. The wavelets are called *bot* (knots or nodes), and a series of them are idealized as being along the north–south line called *okar* (the root)—just as one follows the root to get to the tree, following the okar will lead to the island. The direction that should be taken along the okar is determined by the change in the angles formed by crossing swells; the angles decrease for bot closer to the atoll.

In addition to this use of swell intersections, the bent arms of the swells themselves are used to delimit the sectors of sail. The bent arms of rilib (coming from the east) as it passes the island are termed *rolok* (the northern arm—RM on Figure 4.6c) and *nit in kot* (the southern arm—KS on Figure 4.6c). The former translates into "something lost," that is, you have missed the island, and the latter into "a hole," that is, a cul-de-sac. Both bent arms of kaelib (coming from the west) as it passes the island (TM and KQ on Figure 4.6c) are called *jur in okme* meaning "stakes." When coming toward the center atoll from the north, one should stay between rolok and jur in okme until the okar is found and followed. Similarly, from the south, one should be between nit in kot and jur in okme until finding the okar.

Other meanings for the lines on Figure 4.6c focus on atolls at the points we have labeled 1, 2, 3, 4, rather than in the center. For these, the swells are shown as wedges rather than arcs.

With the wind from the east, for an atoll at 1, FP represents the southern arm of rilib (AB) and GP the northern arm of kaelib, and so 1P is the okar for the atoll at 1 when approaching from the south. Similarly, for the atoll at 2, PH is the northern arm of rilib, PN the southern arm of kaelib, with 2P the okar for the atoll. If there is no atoll at the center between 1 and 2, one should follow the okar from 2 until meeting the okar for 1. And for the atoll at 4, G4N is the bent rilib, while, similarly, F3H is the bent kaelib.

Other descriptions rely on a somewhat different denotation of the words "nit in kot." Rather than the expression referring to just the one demarcation line KS, some use it to refer to the entire region (between RM and KS) on the leeward side of the island in which there is greatly lessened swell energy or, alternately, to both lines that bound this region. In either case, for a wind from the east, when the island is to the east, one should sail north or south to meet one of these boundary lines in order to get out of being a "trapped bird" or, in our terms, having to sail too close to the wind.

On the windward side of an atoll, wave reflection plays the prominent role. For a wind from the east, land in the westerly direction is signaled by observing the reflected bent swells coming back against the main incoming swell. This interaction is said to be observable in a quadrant extending outward from the atoll for about 32 km. Look again at sector AB on Figure 4.12 and the sectors marked, in particular, on Figures 4.7a and 4.7c. For Figures 4.7a–4.7c, now also visualize the curved swell in motion, moving *eastward* to meet the incoming straight swell. In one account, the Marshall Islands navigator says that when one is at the meeting of the reflected and incoming swells, one should put the boat at the corner of the crossing and head away from the corner at an angle that is the same as the swells are forming (see Figure 4.13).

Another description notes that directly windward of the atoll, the reflected swells are more or less parallel to the incoming swells, but at about 45° to either side, the effect of the crossing swells becomes clearly observable. This effect marks the delimiting arms of the quadrant, and so the navigator seeks to sail parallel to one of the arms in order to go directly to the atoll. Both sets of instructions are, in effect, the same: it is when the incoming and reflected swells meet at a 45° angle that heading away from the corner at that angle leads to the atoll

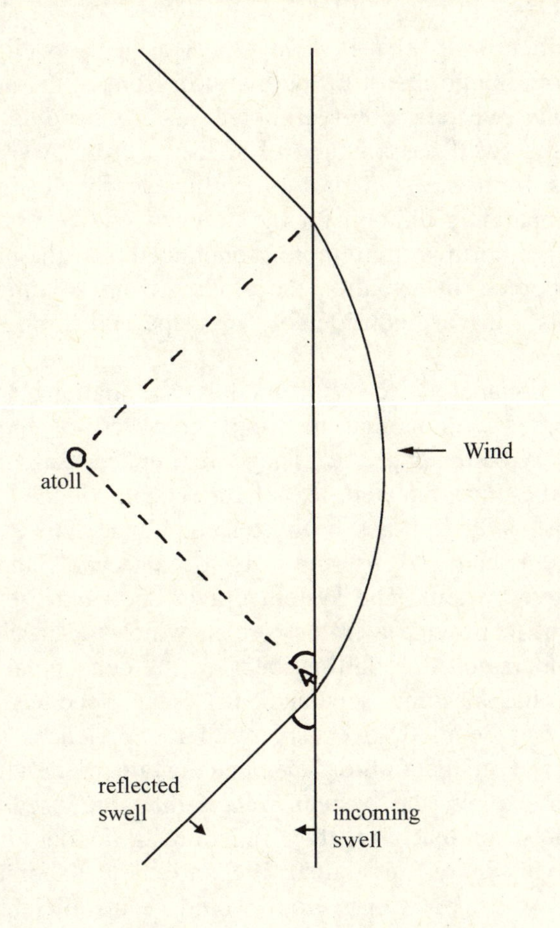

Figure 4.13 On the windward side of an atoll.

and, simultaneously, goes along one of the delimiting arms of the quadrant (again, see Figure 4.13).

Even without further definitive statements about what many details of the mattang represent, it is clear that they are used as models of swell interaction for navigational purposes. They isolate and idealize the swells, emphasizing directionality with respect to wind and land positions. That they are abstract models is highlighted by the use of four uniformly depicted swells from perpendicular directions and land masses symmetrically placed and reduced to points. In actuality, there is a prominent swell that is always present. This swell moves, in our terms, from the northeast to southwest in front of the northeast trade winds. One Western observer noted that, depending on the wind velocity, the swell has a height of at least 1.5 m. A secondary swell pattern is from the southeast to northwest. Swells from the southwest are apparently difficult for the unpracticed eye to observe, and those from the northwest are more pronounced near the northerly atolls. And of course, in actuality, the atolls are *not* symmetrically positioned points, but vary considerably in shape and in size and in relative positions.

As with our explanatory scientific models, the mattang are quite distinct from representations that are simply intended to "look like" or evoke what is being depicted. They are neither seascapes nor maps. Here, the accumulated experiential knowledge of the Marshall Islands navigators has been conceptualized into a general system, a system believed to apply to oceans and landmasses anywhere and everywhere. The system draws upon experience but goes beyond that to provide a framework in which the past experiences can be understood. In addition, because the conceptual system is general and has become separated from the specifics of the experiences, it can be used to consider other experiences that are hypothetical or yet in the future. The mattang are models used to encapsulate and explain the system. When they are used by the Marshall Islanders for teaching, they function as do our diagrams on blackboards or figures in a text. We and they elaborate such depictions with words, but words alone would be insufficient. Particularly for dynamic systems, diagrams play a crucial role. They not only provide a way to visualize the interrelationships of the parts, but enable us to keep the entire system in mind while mentally manipulating or focusing on some part of it.

The essence of an explanatory model is its simplicity and parsimony. It strips the system to what is considered essential. Here, those essentials are phrased in terms of the geometric characteristics of the ocean phenomena—the substances of the land and sea and wind are recast into points, lines, curves, and angles, and the interplay of the phenomena is recast into how these geometric aspects change and interact. Since it is so well stated, to emphasize further the nature of explanatory models, we borrow from Leo Apostel's discussion of them in *The Concept and the Role of the Model in Mathematics and Natural and Social Sciences*:

> ...when a picture, a drawing, a diagram is called a model for a physical system, it is for the same reason that a formal set of postulates is called a model for a physical system. This reason can be indicated in one word: simplification. The mind needs in one act to have an overview of the essential characteristics of a domain; therefore the domain is represented either by a set of equations, or by a picture or by a diagram. The mind needs to see the system in opposition and distinction to all others; therefore the separation of the system from others is made more complete than it is in reality. The system is viewed from a certain scale; details that are too microscopical or too global are of no interest to us. Therefore they are left out. The system is known or controlled within certain limits of approximation. Therefore effects that do not reach this level of approximation are neglected. The system is studied with a certain purpose in mind; everything that does not affect this purpose is eliminated.

Here, we are extending his grouping of "a picture, a drawing, a diagram" to include these artifacts made of palm ribs.

Before we leave the mattang and return to the charts categorized as meddo and rebbelith, a special mention must be made about angles and their measurement. Angles between swells and, in particular, angles that are the same or are increasing or decreasing play a significant role in the system. In practice, these must be determined by the navigator from within the boat and even in the dark of night. A primary method is to lie down in the bottom of the boat and *feel* the rocking from side to side. During their training, navigators are taught to analyze and interpret this kinesthetic information. In fact, one navigator recounted how, as an early part of his training, he was made to float in the water at various places in order to learn how to feel what would later be shown and explained to him.

4 Now, with some understanding of wave piloting and of what is seen by the Marshall Island navigators as features of the environment that are significant for navigation, we can return to the rebbelith and meddo. The charts called rebbelith are maps of the entire archipelago or of one or the other of the atoll chains within it. The meddo are maps of smaller regions. These maps differ considerably from our nautical maps, which emphasize the outlines of land, depths to the ocean floor, and any man-made objects that have been expressly anchored in the ocean or on land to aid mariners. On the stick charts, there are no semi-realistic renderings of indentations along coastlines or promontories, but, rather, large expanses of open ocean marked with lines and curves whose interpretations were introduced with the mattang. On these maps, however, the lines and curves are the *actual*

result of the wind and sea interaction in and around a group of *real* atolls, which vary in size, shape, and underwater topography. Among the maps, even of the same region, there are decided differences. This, in part, is because the maps differ in their levels of generality. In addition, they are products of different schools, based on different atolls and with slight differences in their traditions and styles. And of course, they are handmade objects, made by different individuals.

Figure 4.14a is the same rebbelith as shown in Figure 4.3a, and Figure 4.14b is another rebbelith. On Figure 4.14, however, we have included labels identifying the atolls that are said to be marked by the shells that remain. Also, the rebbelith are now both oriented in the same direction. The orientation is in keeping with the orientation we used in our illustrations of the mattang (Figures 4.5–4.7). In each, the straight edges that are placed vertically are the swell rilib, which is driven by the prevailing wind and the opposing swell kaelib. As we mentioned previously, while the prevailing wind is generally referred to as coming from the east, that is *not* its direction in the Western system of reference. The direction recorded by Westerner observers is approximately 20° north of east. Thus, these charts can be compared to Map 4.2 by rotating the chart illustrations by about 20° counterclockwise. In

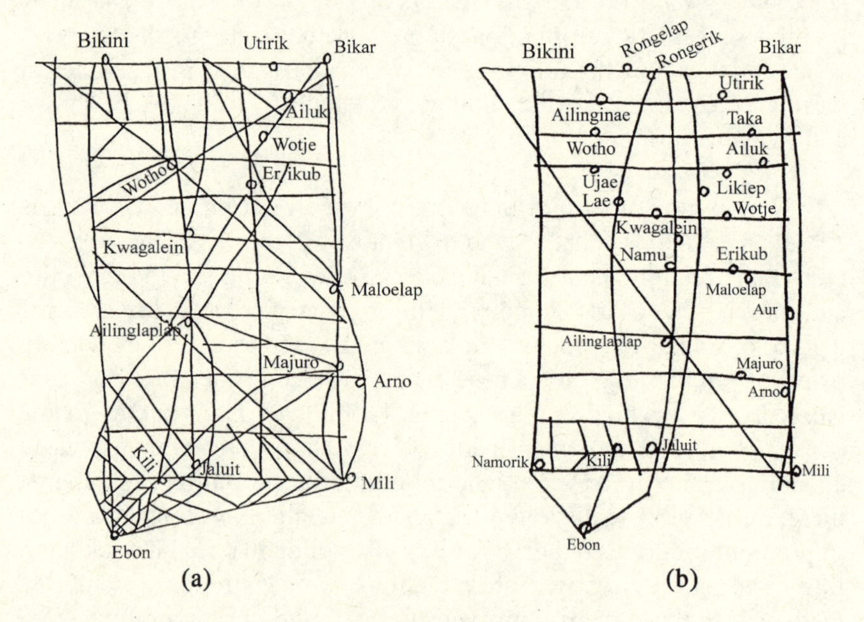

Figure 4.14 Rebbelith.

comparing them, it is important to keep in mind that the region depicted covers over 750,000 sq. km. The distances between Bikini and Rongelap or between Ailinglaplap and Jaluit, for example, are each about 130 km. The positions of the atolls relative to each other are remarkably accurate, given that they were determined without technological aids and that vast open ocean spaces are involved. We can, of course, make no comparison of the swell positions or the positions of the atolls relative to them, as the swells are not marked on our map.

Although Figure 4.14b contains far fewer swells, both charts contain several repetitions of the inward bending rilib shown as arcs and as wedges. Another feature that they share is a line from Kili to Ebon, said to be the okar or direct route between them. Also, on each of the charts, there is a prominent angle formed between the straight rilib and a northeasterly line emanating from just below Mili. While the angles shown are not equal to each other, they are striking in their similarity to each other, and to some of the prominent angles on the mattang.

Figures 4.15a, 4.15b, 4.16a, and 4.16b are meddo. Although they differ, all four of them are renderings of the same subregion. (Figure 4.17 is an actual photograph of the artifact drawn in Figure 4.16a.) All of these meddo are, essentially, blowups of the region in the lower portion of the rebellith shown in Figure 4.14, focusing on the locale of five, six, or

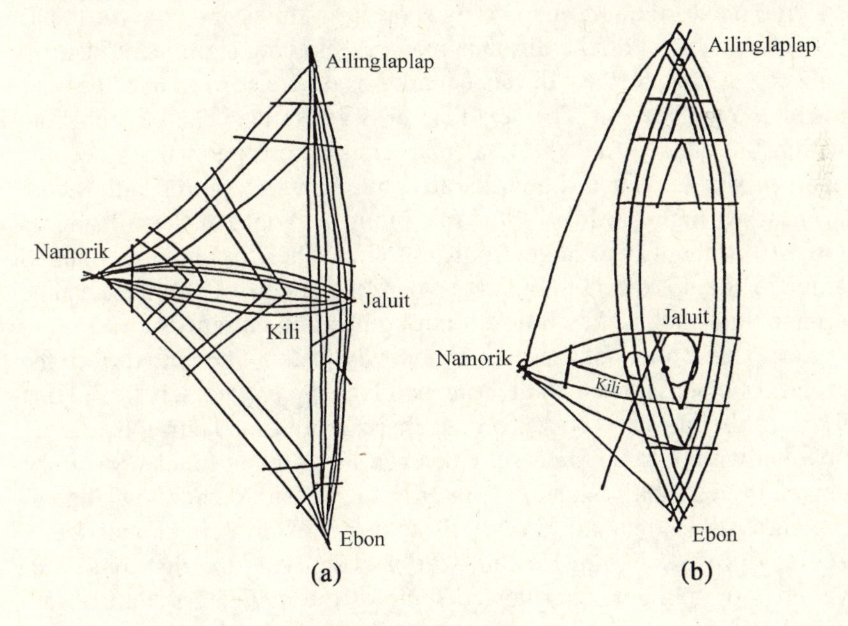

Figure 4.15 Meddo.

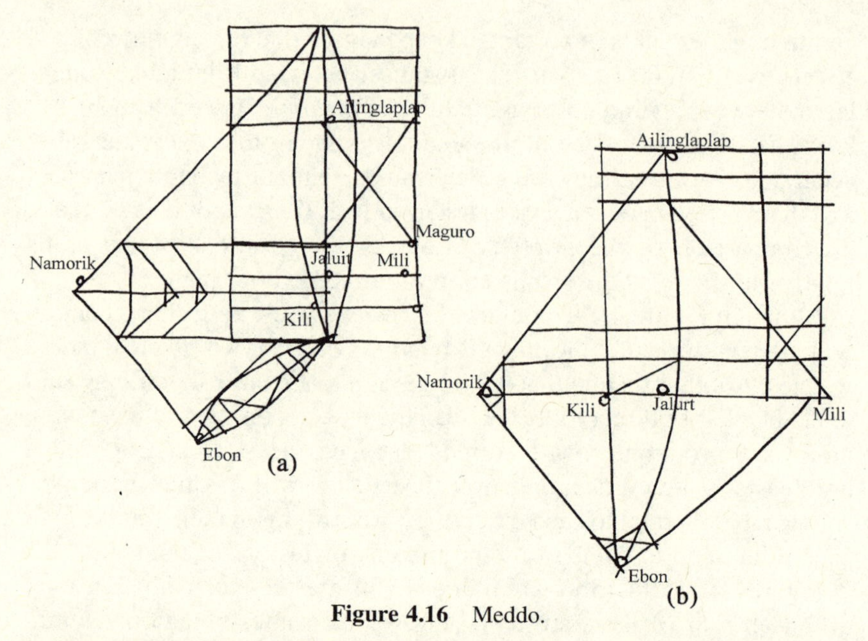

Figure 4.16 Meddo.

seven of the atolls. In addition to the atolls and the ever-present rilib and kaelib, there are other details included on these more localized maps. On these, on the approaches to the atolls along the okar, there are also some specific places that are marked by short cross lines and chevrons. The cross lines are said to be distance markers showing significant identifiable places. The furthest distance marker is where an atoll just becomes visible on the horizon. The next marker is where land can be seen from within the canoe. And the final marker is where palm trees become distinguishable. Since the atolls are quite low, averaging only about 3.7 m above high-tide level with a maximum of about 9 m, these distances range from about 16 to 24 km from an atoll. Further, the chevrons marked along the okar are said to indicate where the effects of ebb tides can be seen in the water as they flow out through lagoon entrances.

Early interpreters of the stick charts, who had not benefited from the reports of discussions with the Marshall Island navigators, believed that many of the lines and curves on the charts represented currents. This is now known not to be the case. However, although not included on the charts, the Marshall Island navigators have considerable knowledge of the currents in their area. In 1949, to learn from their knowledge, a zoology professor from Hawaii, who was studying the distribution of varieties of sponges, interviewed some of the navigators. Part of this discussion focused on the vicinity depicted on the meddo in Figures

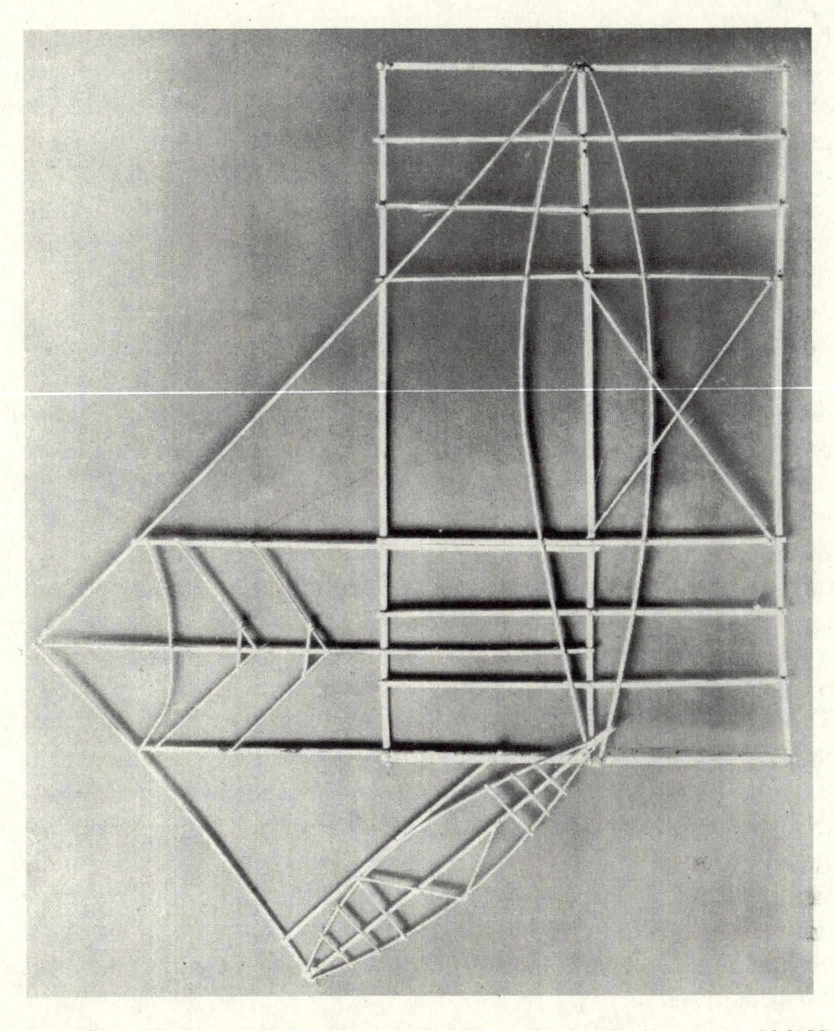

Figure 4.17 A photograph of the meddo in Figure 4.16a. (Catalogue No. 206188, Department of Anthropology, Smithsonian Institution.)

4.15 and 4.16. There is, they maintain, a quiet current-free sea south of Ailinglaplap. This, they say is because there is a shoal and, therefore, an area with comparatively shallow water. As a result, in addition to being free of current, the region reacts to tide changes in quite the same way as does a lagoon inside an atoll. That is, there is a channel through which the rising tide streams in and the falling tide rushes out. This anomalous region is south of Ailinglaplap extending to a line between Namorik and Jaluit, and its channel is east of Namorik between

Namorik and Kili. The chevrons on the meddo between Namorik and Kili might well be related to the effects of the ebb tides through this channel. In any case, this anomalous region is seen to be specially prominent on the rebbelilth in Figure 4.14, as well as on the meddo in Figures 4.15–4.17.

As with the mattang, the stick charts that are maps are used on land for the preservation and transmission of knowledge. The information that the navigators learn from them is carried on a voyage, but the objects themselves are not. Reliance on memory is often found in oral cultures such as the Marshall Islands culture at the time the charts were made and used. Those of us in literate cultures, who have come to depend on the media of literacy, are often both surprised and impressed by what others are able to commit to memory. For the Marshall Islanders, the artifacts from which they learned are apparently not needed while at sea. Their contents, or the cues they provide, are committed to memory, supported by other devices that are also learned as part of navigational training. One such set of devices are *rojen kōklōl* (translated as "navigational formulas" or "indicator mnemonics"). These are used to remember sailing directions and indicator signs for particular routes. They are said also to have magical properties that help maintain a high level of confidence during a difficult voyage. Closely related to these are *alinlōkōnwa* (translated as "sailing songs" or "songs from the stern sheets"). These are sung by the canoe steersmen, and although they are relatively short, each can be repeated for many hours. Their function, too, is to maintain alertness and confidence, while reminding the navigator of dangers and indicators. In addition, the number of times a particular song is repeated is used as an aid in measuring elapsed time and, hence, assists in estimating how much of the course has been covered.

5 As with other representations of knowledge among traditional peoples, we are limited in our ability to read the stick charts in full. Not only do we not share in the Marshall Islands culture, but we lack the specific training and the shared experiences of the navigators. We can, nevertheless, recognize that the rebbelith and meddo are planar representations of particular regions in and around the Marshall Islands, including what is significant about the regions for the navigators. As such, they are expressions of the most important mapping concepts. As with all maps, the creators of the stick charts that are maps have drawn upon diverse information and a shared navigational tradition to create

analogical spaces. Within the spaces, physical entities viewed as significant in their navigational tradition are symbolically represented, and, most important, relations that may not be seen directly while in the original spaces are made visually explicit in the analogical spaces.

The mapping concepts evidenced by the rebbelith and meddo are important. However, it is the mattang and the mathematical and scientific ideas they embody that we consider even more substantial and more significant. As models of the conceptual framework underlying the wave piloting system, they not only attest to the fact that such a framework has been created, but encapsulate the conceptual framework in a visual rendering. Abstraction, generalization, and idealization all come into play.

In the history of modern science, understanding of the physical world and mathematical modeling have been intimately connected. In fact, geometric and then algebraic representations of physical systems have been the hallmark of modern science. The linkage is so tight that it is hard to conceive of the study of physics without the involvement of mathematical ideas and mathematical descriptors. Why this is so, or why this should be so, has long been the subject of philosophical discussions. These discussions have included speculation on whether this is the nature of the universe, or the nature of the human mind, or simply the scientific tradition of Western culture. Because it is outside the modern tradition, the Marshall Islands case offers an unusual contribution that may enrich these discussions.

For this case also, broadly dispersed experience and observations of natural phenomena are given meaning by being conceptualized into a coherent, structured system. The system is represented via a model that simplifies, highlights, and isolates what is considered essential. Most important, what is selected as essential are the geometric aspects of the phenomena. Here, the natural phenomena are oceanographic; the system is a conceptualization of the interplay of land, sea, and wind; the model is a planar representation. Relative positions and relative directions become all important. Swells are reduced to curved lines, and there is attention given to the locus of points where the curves cross, as well as to the changing crossing angles as the curvature changes. In short, the system becomes one characterized by geometric essentials that give rise to geometric implications. And, what is more, the compact planar representation becomes the vehicle for transmitting to future navigators the accumulated experience subsumed by the conceptual system. A basic tenet of modern science is that different

Figure 4.18 A U.S. postage stamp issued in 1990. The stick chart shown is the rebbelith in Figure 4.14b. (Stamp Design © 1990 U.S. Postal Service. Reproduced with permission. All rights reserved.)

observers, at different places, and at different times, may make different specific observations but, none the less, will find the same principles at work. This premise, too, is present as it clearly underpins the creation, transmission, and use of this conceptual system.

Made of palm ribs, coconut fibers, and shells, the Marshall Island stick charts underscore the fact that analogical planar representations, whether of space or of physical systems and their inner relationships, are quite independent of writing systems and are not confined to any particular culture or any particular medium. Consideration of the stick charts provides us with a deeper understanding of the intellectual endeavors of the Marshall Islanders, but, in addition, it adds to the growing realization of the importance of visualization and representation in the mathematical and natural sciences.

Notes

1. The dictionary definition of a map is quoted from *Webster*'s *New Collegiate Dictionary*, G. & C. Merriam and Co., Springfield, MA, 1973, p. 701. The quotes from Ptolemy are from p. 61 of Lloyd A. Brown's, *The Story of Maps*, originally published in 1949 and reprinted in 1977 by Dover Publications, New York. The book is a very readable overview of the history of Western maps. It discusses maps and mapmaking as closely tied to exploration, trade, and colonialism. World War II was another major impetus for interest in maps. Also cited in the book are the contributions to mapmaking of numerous mathematicians, such as Picard, Airy,

Jacobi, Laplace, Liebnitz, and Newton. Highly recommended for its broad consideration of maps and mapping is *The Nature of Maps: Essays in Understanding Maps and Mapping*, Arthur H. Robinson and Barbara B. Petchenik, University of Chicago Press, Chicago, 1976. Some interesting insights about maps and map-reading are on pages 149–153 of "The emergence of a visual language for geological science 1760–1840," Martin J.S. Rudwick, *History of Science* 14 (1976) 149–195. *Maps are Territories: Science is an Atlas*, David Turnbull (with a contribution by Helen Watson with the Yolngu community at Yirrkala), and *Singing the Land, Signing the Land*, Helen Watson (with the Yolngu at Yirrkala) and David Wade Chambers, both published by Deakin University Press, Gelong, Victoria, Australia, 1989, contain excellent ideas to help broaden the way we think about maps. They then present as maps bark paintings of a native Australian group. A recommended discussion of Inuit maps is "A cultural interpretation of Inuit map accuracy," Robert A. Rundstrom, *The Geographical Review*, 80 (1990) 155–168. Also, an extensive collection of Inuit maps is in *Eskimo Maps for the Canadian Eastern Arctic: Cartographica Monograph No. 5*, John Spink and D.W. Moodie, 1972. A discussion of the spatial concepts of the Inuit is found on pp. 132–140 in my *Ethnomathematics: A Multicultural View of Mathematical Ideas*, Chapman & Hall/CRC, New York, 1994.

It is interesting to note that grid systems on maps were used in China in the second or third century CE. For more about this, see pp. 30–33 in *The Genius of China: 3000 Years of Science, Discovery, and Invention*, Robert Temple, Simon & Schuster, New York, 1986.

2. For further reading on the environment and lifeways of the Marshall Islanders, see *Majuro: A Village in the Marshall Islands*, Alexander Spoehr, Fieldiana: Anthropology series, vol. 39, Chicago Natural History Museum, 1949 (Krauss Reprint Corp., New York, 1966); Leonard Mason, "Suprafamilial authority and economic process in Micronesian atolls," chapter 16, pp. 299–329 in *Peoples and Cultures of the Pacific*, Andrew P. Vayda, ed., Natural History Press, New York, 1968; and chapters 1–3 of *The Bikinians: A Study in Forced Migration* by Robert C. Kiste, Cummings Publishing Co., Menlo Park, CA, 1974. (According to an article in the *International Herald Tribune* [May 21, 1996, p. 7, cols. 5 and 6] the fame of Bikini atoll was spread by the scant swimsuit named for it by the French designer Louis Réard at the time of the atomic testing.)

The women's dress mats are rectangular with sometimes as many as seven different patterns around the periphery. About fifteen mats are shown in Plates 9, 10, and 11 of *Ralik-Ratak (Marshall-Inseln)*, Augustin Krämer and Hans Nevermann, Ergebnisse de Südsee-Expedition 1908–1910, G. Thilenius, ed., Part II B, vol. 11, Friederichsen, de Gruyter & Co., Hamburg, 1938. Discussion and analysis of strip decorations in general, and those of the Maori of New Zealand and Inca of South America in particular, are in chapter 6 (pp. 166–183) of *Ethnomathematics*, cited above in the notes to section 1.

3. Much of the material in this chapter, particularly in this section and the next, is adapted from my article "Models and maps from the Marshall Islands: A case in ethnomathematics," *Historia Mathematica*, 22 (1995) 347–370. See that article for more specific citations and additional references. The article also contains an appendix that may be of special interest to those who wish to pursue further study of the charts. The appendix summarizes the widespread corpus of stick chart illustrations.

In all, it cites 103 published illustrations of sixty-nine different charts, including notes on which illustrations are photos, which are drawings, and which appear to be similar to each other, although they are different artifacts.

Some useful readings for understanding oceanographic phenomena are Willard Bascom's *Waves and Beaches: The Dynamics of the Ocean Surface*, Science Study Series, Doubleday & Co., New York, 1964; *Wind Waves*, Blair Kinsman, Dover Publications, New York, 1984, pp. 156–167; and chapter 5 in *Waves and Tides*, R.C.H. Russell and D.H. Macmillan, Hutchinson's Scientific and Technical Publications, London, 1952.

Numerous books are available as an introduction to sailing in Euro-American sailboats. See, for example, *Invitation to Sailing*, Alan Brown, Simon & Schuster, New York, 1968, or chapter 2 in *The Lure of Sailing*, Everett A. Pearson, Harper and Row, New York, 1965. A discussion of the outrigger canoe and how it handles under sail is in chapter 3 of Thomas Gladwin's *East is a Big Bird: Navigation and Logic on Puluwat Atoll*, Harvard University Press, Cambridge, MA, 1970. The entire book is highly recommended, although it is about a distinctly different navigational tradition; namely, that of the Caroline Islands navigators.

A very good discussion of models is in Ian G. Barbour's *Myths, Models and Paradigms: A Comparative Study in Science and Religion*, Harper and Row, New York, 1974 (see, in particular, pp. 6–7 and 29–38). Although most of the book is about computer simulation, the beginning portions of *Would-Be Worlds: How Simulation is Changing the Frontiers of Science*, John L. Casti, John Wiley & Sons, New York, 1997 are devoted to models in general. Most important, however, is Leo Apostel's "Toward the formal study of models in the non-formal sciences," pp. 1–37 in *The Concept and Role of the Model in Mathematics and Natural and Social Sciences*, International Union of History and Philosophy of Sciences: Division of Philosophy of Sciences, D. Reidel, Dordrecht, 1961. The portion quoted in this section is from p. 15 of his article. It is quoted with the kind permission of Kluwer Academic Publishers.

Valuable discussions of the interpretation and use of stick charts are found in William H. Davenport's "Marshall Islands Navigational Charts," *Imago Mundi*, 15 (1960) 19–26, and his "Marshall Islands Cartography," *Expedition* (Bulletin of the Museum of University of Pennsylvania), 6 (1964) 10–13; M. W. de Laubenfels, "Native Navigators," *Research Review* (Office of Naval Research), June 1950, pp. 7–12; David Lewis, *We, the Navigators*, University Press of Hawaii, Honolulu, 1972; Captain Winkler, "On sea charts formerly used in the Marshall Islands, with notices on the navigation of these Islanders in general," pp. 487–508 in *Annual Report of the Smithsonian Institution for the Year Ending June 30, 1899*, Washington, DC; and the book by Krämer and Nevermann included in the notes to section 2. (The quotation from the navigators about the general applicability of the mattang is from p.10 of de Laubenfels' article.)

The mattang in Figure 4.5 is in the collection of the Smithsonian Institution, Department of Anthropology, Catalogue No. 206187. It was photographed from the article by Winkler cited above (Plate VI).

In 1902, A. Schück published a compendium of drawings of all Marshall Islands stick charts that he had located in museums and private collections. Several of my figures follow his drawings. The mattang in Figures 4.6a–4.6c, are after Schück's figure 11, Winkler's Plate XIIIa, and Davenport's figure 3, respectively. Those in Figures 4.7a–4.7e, are after Schück's figure 7; figure 2 in Henry Lyons' "The Sailing Charts of the Marshall Islanders," *The Geographical Journal*, 72 (1928) 325–328; Plate III in N.O. Hines' "The Secret of the Marshallese Sticks," *Pacific Discovery*, 5 (1952) 18–23; and Schück's figures 39 and 15, respectively.

4. For some references to discussions of the interpretation and use of the stick charts, see the notes above for Section 3. The discussion of the current-free sea south of Ailinglaplap is in M.W. de Laubenfels' "Ocean currents in the Marshall Islands," *Geographic Review*, 40 (1950) 254–259. The navigational formulas and sailing songs are discussed in William H. Davenport's "Marshallese folklore types," *Journal of American Folklore*, 66 (1953) 219–237.

The rebbelith in Figures 4.14a,b, are after Schück's figure 27 and Winkler's chart II. (Figure 4.3a is the same as Figure 4.14a but without labels.) The meddo in Figures 4.15a and 4.15b are after Schück's figures 43 and 42, and Figure 4.16b is after his figure 36. Figures 4.16a and 4.3b are drawings of the meddo in the photo in Figure 4.17.

5. The linkage between mathematical modeling and understanding of the physical world is the major theme of the well-known *Mathematics and the Physical World*, Morris Kline, Anchor Books, New York, 1963. His chapters 1 and 2 contain a general overview, which is then substantiated throughout the rest of the book. See also his summary chapter 27.

The acknowledgment of the importance of visualization and representation in the sciences and in mathematics and mathematics education has been stimulated by the advent of computer technology. Although they were static, well-conceived diagrams, drawn on paper or blackboards, have always played a crucial role in texts, classrooms, and professional presentations. An important re-examination of the role of diagrams in Greek mathematics is *The Shaping of Deduction in Greek Mathematics: A Study in Cognitive History*, Reviel Netz, Cambridge University Press, Cambridge, 1999. See also his "Greek mathematical diagrams: Their use and their meaning," *For the Learning of Mathematics*, 18 (1998) 33–39.

CHAPTER 5

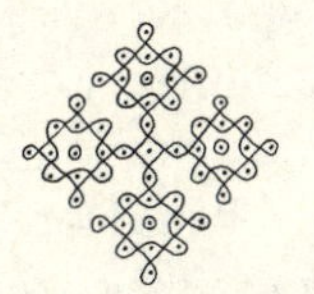

Systems of Relationships

1 Relations are fundamental to mathematics and mathematical structures. Any specified property linking a pair of objects is a relation. "Less than," "more than," "four times as much as," "equal," or "unequal" are properties that apply to number pairs, but other relations, such as "taller than," "lighter than," "older than," "sister of," in addition to "equal" or "unequal," are properties that can apply to many different things, as well as to people. In a relation such as "taller than," whether applied to objects or to people, a physical attribute is being compared. However, a relation such as "of higher rank" is different in type; it involves a judgment that is solely cultural. It differs from culture to culture and even has no meaning in some cultures.

In mathematical systems, as well as in systems of social organization, there are complexes of relations that involve several relations, sets of items to which they apply, and, moreover, relations among relations. It is their interconnectedness and interdependence that create from the parts an overarching entity that is deemed a system. In describing a system, whether a social or mathematical system, we must specify the set of objects being discussed; we must be clear about what characterizes the relations; and we must be cautious that the relations that make up the system are not conflicting.

Here, we begin with a discussion of a single relation—the equality relation—first exploring its common and mathematical meanings and then enlarging to its meaning and use in a system of social organization among the Basque of Sainte-Engrâce. Then, turning to the Tonga of Polynesia, we view a system of status relations that, in fact, contains an internal contradiction. (We will also see how that contradiction is resolved.) Lastly, our discussion of the logic of relations in systems of social organization will move us to Africa and to the elaborate Gada system of the Borana who live in Ethiopia near its border with Kenya. Throughout, we emphasize that it is the clear and formal statements of these systems *by the people who live them* that enable our discussion. That is, it is not we who are imposing the logic on observed behavior. Rather, it is the people in the cultures who are articulating the properties of the relations that we can then discuss in terms familiar to us.

2 In mathematics, the equality relation is one of several that are formally categorized as *equivalence relations*. An equivalence relation has three specific properties: symmetry, reflexivity, and transitivity. Symmetry means that if a is related to b in some way, then b is related to a in that same way. (With equality of numbers, for example, that means that if $a = b$, then $b = a$. A contrasting example is the relation *greater than* which is not symmetric: if a is greater than b, then b is *not* greater than a.) Reflexivity of a relation means that the relation holds between an element and itself. (Numerical equality does have this property: $a = a$. For greater than, we cannot say "a is greater than a", and so the relation does not have the property.) The third property, transitivity, specifies that if a is related to b in some way, and b is related to c in the same way, then the relation also holds between a and c. (For the equality of numbers, this is true: if $a = b$ and if $b = c$, then $a = c$. For greater than, although the former two properties did not hold, this property does; that is, if a is greater than b, and b is greater than c, than a must be greater than c.)

An equivalence relation, familiar to those who have studied geometry, is the congruence of triangles in the plane: (i) if triangle A is congruent to triangle B, then B is congruent to A; (ii) triangle A is congruent to itself; and (iii) if triangle A is congruent to triangle B, and triangle B is congruent to triangle C, then triangle A is congruent to triangle C. And another equivalence relation has already been encountered and used in previous chapters; it is the one involved in modular arithmetic. We used, for example, the modular equivalence $15 = 3 (\text{mod } 4)$. In this relation, the equal sign does not have the same exact meaning as in our ordinary numerical equality, but the relation represented does share the three properties defining an equivalence relation.

We often pass easily from one form of mathematical equivalence to another, but, beyond numerical equality, which we have incorporated from our earliest learning, we do have to be cautious as to what specifically is meant. Each time a new set of mathematical objects is encountered (e.g., sets, groups, modular numbers, complex numbers, matrices, or vectors), we must clarify what is meant by two of the objects being equal; that is, we must specify a new equivalence relation.

In common American-English usage, the connotations of *equality* and *equivalence* differ: equality "implies the absence of any difference," that is *exactly* the same, while equivalence implies that, although there may be differences, "they amount to the same thing." For the political realm, where equality is so central to the Euro-American views of democracy, justice, and fairness, equality is used mainly in regard to the rights or treatment of people vis-à-vis government, institutions, or businesses. There is a long and ongoing history of philosophical and legal discussions of equality. Still, however, when used in the sociopolitical realm, there are deep and significant disagreements with even common catchphrases meaning different things to different people. For example, in recent United States history, the phrases "separate but equal," "equal pay for equal work," and "equal opportunity" have been used, argued, interpreted and reinterpreted.

Among the Basque of Sainte-Engrâce, France, there is, in the realm of social organization, a concept *bardin–bardina* translated into English as "equal–equal." This concept of equality is markedly different from ours. In contrast to our mathematical or sociopolitical ideas of equality, the Basque concept is intertwined with the idea of circularity and is dynamic rather than static. Its focus is on systematizing the interactions of the group members in order to provide mutual assistance and receive mutual benefit. Examining their concept of equality enlarges our view of this seemingly simple word.

3 The community of Sainte-Engrâce is in the Basque province of Soule, one of the nine Basque provinces in the Pyrénées-Atlantique which straddle the French–Spanish border. Although the exact origins of the Basque are unknown, it is generally agreed that they predate the French- and Spanish-speaking peoples in the region around them by, perhaps, thousands of years. In the 1970s, at the time of a study of Sainte-Engrâce, there were about two million Basque, with about three-quarters of them living in the Basque provinces in Spain, one-eighth in the Basque provinces in France, and the rest living in other areas of the world. Having their own language, a rich history,

their own political and social organization, and long-held traditions, the recent history of the Basque has been marked by conflict with the nation states that encompass them. Nevertheless, the Basque way of life continues, particularly in a place like Sainte-Engrâce, which, situated in the high mountains, is one of the most geographically and socially isolated communities in the region. Although the population declined from about 1000 people in the late 1800s to about 375 in the 1970s, the community remains self-reliant, centering on small farms and shepherding.

The mountains that surround the Sainte-Engrâce region range from about 1000 to 2500 m. The Basque conceive of the region in which they live as enclosed by a circle of mountains with their households forming another circle within that. Whether or not this is actually the case, this spatial model forms the basis for their idea of circularity that pervades many of their interactions. In this circle, everyone has neighbors to the left and neighbors to the right. No one is first, and no one is last. Everyone's participation is involved in keeping the circle unbroken.

The Basque concept of equality is underpinned by two operational principles that structure relationships so that everyone both gives and receives. The principles are referred to as *üngürü* and *aldikatzia*. The former is translated into English as "rotation," in the sense of "moving around a centre," and the latter as "'serial replacement" as well as "alternation." How these mathematical ideas apply in this context and how they relate to equality are best described in terms of their operation.

A fundamental circular exchange, which was in effect until the 1960s, was the giving of blessed bread. Each household regards its neighbor to the right as its *first* neighbor. (The directions right and left are as viewed from the center of the circle so that right is clockwise and left is counterclockwise.) The giving of bread took place weekly and was thought of as being given from first neighbor to first neighbor. That is, each Sunday, a woman from one particular household brought two loaves of bread to the church where it was blessed and partially used in a church ritual. Then, before sunset, a portion of the bread was given by her to her first neighbor. The following week the first neighbor was the bread giver, and *her* first neighbor was the bread receiver. Thus, the giving (and receiving) of bread moved around the circle serially, taking about two years to complete one cycle of about 100 households. While each household was both a giver and receiver of bread, this mode differed from simple reciprocity; only if there were a total of two households would neighbors directly reciprocate.

To describe more succinctly the foregoing exchange as well as those to come, we identify the households in the circle as H_1, H_2, ..., H_n where the numbers in the subscripts reflect the position of the households in the circle, and n is the total number of households. Thus, if some H_i (where $i = 1$, or 2, or 3, ... or n) is the bread-giver on a particular Sunday, she gives to her first neighbor, H_{i+1}. The next week, H_{i+1} gives to H_{i+2}. Because the households are in a circle, the arithmetic in the subscripts is mod n. If, for example, there were, in all, only five households, the circle would contain H_1, H_2, H_3, H_4, H_5, and on H_5's right is H_1, her first neighbor; that is $H_{5+1 (\text{mod } 5)} = H_1$. (Modular arithmetic was introduced in Section 1.2 and then used again in Chapter 3.)

In a more extensive, ongoing, cooperative arrangement, the exchange among neighbors relies on the same circular model, but this exchange involves several first neighbors. The *first* first neighbor of H_i is, as in the bread-giving, H_{i+1}, the neighbor to the right; the *second* first neighbor of H_i is the neighbor on the left (H_{i-1}); and the *third* first neighbor is the next on the left (H_{i-2}). Thus, for example, when there is a death in household H_i, the household calls upon its first neighbors for assistance. As a group, H_{i-2}, H_{i-1}, and H_{i+1} help to keep the household going, but H_{i+1} provides particular assistance in specific preparations for the funeral. And, on the occasion of a home birth for H_i, it is a woman of household H_{i-1} who serves as the midwife (see Figure 5.1).

Planting, harvesting, threshing, sheep shearing, and pig slaughtering all require the work of more than one person, and so, there too, the first neighbors are called upon. These assistances are directly reciprocated by providing food and drink and by the giving of small gifts, but, primarily, the reciprocation is serial, that is, by assisting, when called upon, as the first neighbors of others.

A particularly interesting result of this mode of interaction in the farming yearly round is that households must schedule their work with the obligations of others, and to others, in mind. Also, for the same chore, each household gets to work with different groups of households and to play different roles within those groups. H_i, for example, works in groups (H_{i-2}, H_{i-1}, H_i, H_{i+1}), (H_{i-3}, H_{i-2}, H_{i-1}, H_i), (H_{i-1}, H_i, H_{i+1}, H_{i+2}), and (H_i, H_{i+1}, H_{i+2}, H_{i+3}), taking the roles of primary household, and first, second, and third first neighbors, respectively. And, to avoid causing conflicting obligations for himself or any of his neighbors, H_i cannot schedule his household's work on the same day as the work of

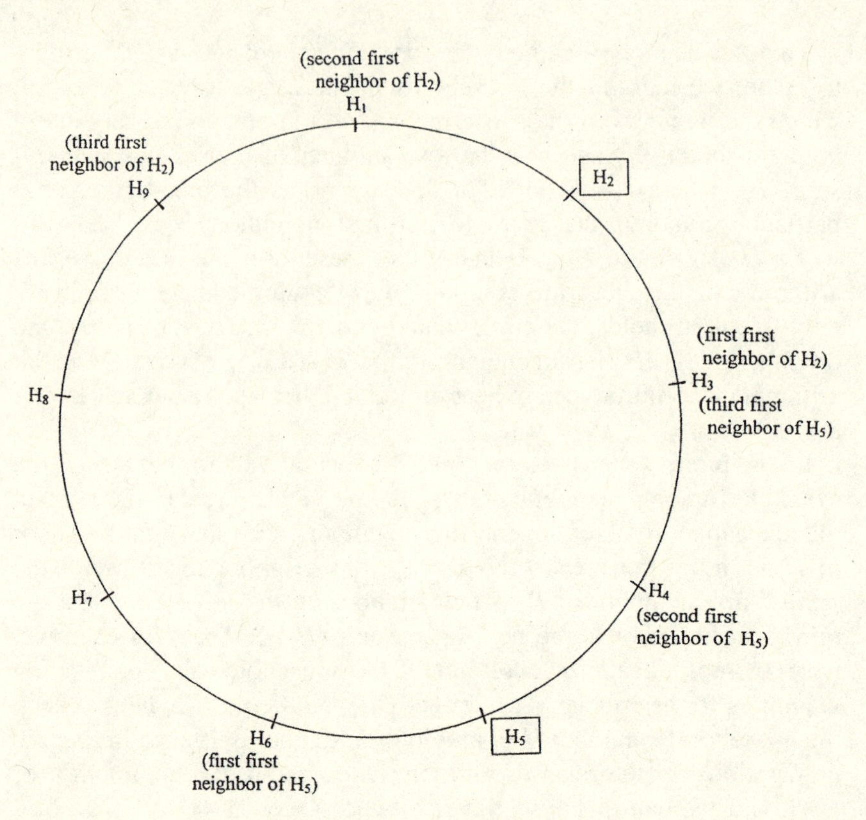

Figure 5.1 The first neighbors of H_5 and H_2 where the number of households is nine.

H_{i-3}, H_{i-2}, H_{i-1}, H_{i+1}, H_{i+2}, or H_{i+3} because, for example, H_i's third first neighbor (H_{i-2}) is H_{i-3}'s first first neighbor, and his first first neighbor (H_{i+1}) is H_{i+3}'s second first neighbor.

By far, the most intricate cooperative arrangement involves the shepherding and cheese-making groups that work and live together during the summer months. These groups of households share in the ownership of pasturage sites in the mountains. The origin and practices of these groups are part of a long tradition that was described in writing as early as the 1600s. Prior to the 1900s, the ideal ownership group consisted of 10 households, each contributing 50 to 60 ewes and 2 rams to the summer flock and one man to the working unit. The flock of about 550 sheep had to be driven up into the mountains in late May, watched over until they were driven down to the valley for shearing in July, then driven back up to be watched over until returning to their valley homes at the end of September. Additional important aspects of the May to

July work were the twice-daily milking of the sheep and the making of cheese from the milk. Different roles were defined that encompassed the various jobs that needed doing, and a formal system of rotation was used to insure that everyone was equal in terms of work contributed, in terms of cheeses produced, and in terms of status.

As contrasted to the first neighbor obligations, which are predicated on the model of households *physically* positioned around the periphery of a circle, the model for the shepherding and cheese-making cooperatives consists of a multiplicity of abstract cycles. The households, first of all, were assigned a specific order in the ownership group that remained unchanged from year to year. For the May through July period, for the working group of ten men, there were six explicit roles which required six of the men to be together at the mountain site. Thus, calling the households' representatives H_1, H_2, ..., H_{10}, and the work roles ranked in status order R_1, R_2, ..., R_6, once the sheep were safely at the mountain site, assuming the household count started with H_1, the assignments were: $H_1 \rightarrow R_1$, $H_2 \rightarrow R_2$, ..., $H_6 \rightarrow R_6$, and H_7, H_8, H_9, H_{10} returned home. After 24 hours, the rotation would begin: H_7 would ascend the mountain, keeping to the right, and then H_1 would descend, keeping to the left. Their ascent and descent are conceived of as taking place in a circle. Upon his arrival on the mountain, H_7 would take on role R_6 and each of the others would move up one role: $H_2 \rightarrow R_1$, $H_3 \rightarrow R_2$, ..., $H_7 \rightarrow R_6$. Similarly, every 24 hours, at the end of the day, there would take place the rotation up and down and the moving up of one role by the others. Thus, on day i, those on the mountain would be H_i, H_{i+1}, H_{i+2}, H_{i+3}, H_{i+4} and H_{i+5} in roles R_1 through R_6, respectively. With ten men cycling through this rotation, the subscript arithmetic is mod 10, and so, more precisely, the assignments are $H_{i(\text{mod } 10)} \rightarrow R_1$, $H_{i+1(\text{mod } 10)} \rightarrow R_2$, $H_{i+2(\text{mod } 10)} \rightarrow R_3$, $H_{i+3(\text{mod } 10)} \rightarrow R_4$, $H_{i+4(\text{mod } 10)} \rightarrow R_5$, $H_{i+5(\text{mod } 10)} \rightarrow R_6$ (see Figure 5.2). On, say, the eighteenth day, those on the mountain would be H_8, H_9, H_{10}, H_1, H_2, and H_3 in roles R_1, R_2, ..., R_6, respectively. Out of every 10-day period, each man spent 6 consecutive days on the mountain and 4 days at home. Generally, from May to mid-July, each of the ten men carried out each of the six roles about six times with, for reasons of equity to be explained later, an extra turn at R_1, for H_1 and H_2.

From the time of shearing in July until the end of September, because milking and cheese- making were complete, the number of men needed at the mountain site was reduced to two with just two roles, which we will call R_1' and R_2'. For this, two men remained on the mountain for 6 consecutive days, alternating daily between roles R_1' and R_2'. After the 6-day period, the pair descended the mountain and the next pair in the cyclic order ascended. Thus, if the period began with H_1 and H_2 in roles R_1' and R_2', then the next day $H_2 \rightarrow R_1'$ and $H_1 \rightarrow R_2'$, and so on until, on the 7th day, $H_3 \rightarrow R_1'$, while $H_4 \rightarrow R_2'$. Since there is a cycle of ten men, in groups of two, for sets of 6 days, and alternating roles, the general expression for who is on the mountain in what role on the *i*th day is more complicated than the previous generality. In general, on the *i*th day of this second phrase, if i is odd, $H_{(2\lfloor(i-1)/6\rfloor+1)(\mathrm{mod}10)}$ is on the mountain in role R_1'. If i is even, he is in role R_2'. The collection $\lfloor(i-1)/6\rfloor$ represents the value of the greatest integer less than, or equal to, $(i-1)/6$. For example, if $i = 9$, the value is 1; for $i = 23$, it is 3; and for $i = 38$, the value is 6. On these days, the assignments would be H_3 in role R_1'; H_7 in role R_1' and H_3 in role R_2', respectively. Their partners, H_4, H_8, H_4, therefore, are in the other role, that is in R_2', R_2', and R_1'. In all, during a 30-day period, each man spent 6 consecutive days at the mountain site, three of them as R_1', and three as R_2', and 24 days at home. Usually, each man had two of these 6-day turns on the mountain.

By these rotations, the men's contributions were the same in terms of time spent at home, time spent at the mountain site, time spent in each of the six roles $R_1, \ldots, R_6$, and time spent in the roles R_1' and R_2'. The procedure also insured receiving an equal number of cheeses made from the milk of the sheep. These cheeses were an important part of a household's annual food supply. One responsibility that went with the highest status role (R_1) was making two cheeses and watching over the cheeses that others had previously made. With the exception of the first cheese made on the first day and the first cheese made on the second day, the cheeses made by a person were for his household's use during the year. (The first cheese was sold outside of the community with all the members of the group sharing equally in the profit, and the other was given to the priest or guard of the forest. The extra turns noted before of H_1 and H_2 being R_1 and, hence, of making more cheese, were to compensate for these cheeses.) In general, a cheese weighed about 8 or 9 kg. With six turns at being R_1 and making two cheeses on each of these days, each person took home about 100 kg of cheese.

A larger cycle in which the annual cycles are embedded is the multi-year cycle. We noted that the ten cooperating households are in a fixed order—$H_1, H_2, \ldots, H_{10}$. The order remains fixed throughout time, but which household representative starts a year as R_1 advances by one position each year. That is, in a hypothetical Year 1, $H_1, H_2, \ldots, H_6$ are the first subgroup at the mountain site, but then in Year 2, the first subgroup would be $H_2, H_3, \ldots, H_7$, and so on, from year to year. (To reflect this in our previous statements involving H_i, i should be modified to $i + Y - 1$ where Y is the year number of the cooperative's operation.)

Finally, we introduce the crucial issue of equality of status, which becomes particularly significant for groups smaller than the ideal of ten. The six roles, which we have simply been referring to as Rs, from highest to lowest status are: R_1 = woman of the house; R_2 = master shepherd; R_3 = servant shepherd; R_4 = guardian of non-lactating ewes; R_5 = guardian of lambs; and R_6 = female servant. R_1 is the cheese-maker and is also in charge of cooking and cleaning the hut in which the six men live. R_6 functions as his servant in the household chores. R_2, the master shepherd, organizes and directs the work of R_3, R_4, and R_5. Because there is a decided hierarchy in the roles, the rotation is of special significance in preserving equality. Having ten men rotate through the six roles insures that no status hierarchy is consistently imposed. In particular, whoever serves as house servant (R_6) when some H_i is woman of the house (R_1) will serve as woman of the house (R_1) when that H_i is house servant (R_6). And the Basque further note that this H_i will never be above those whom his house servant (R_6) will be above when he serves as woman of the house (R_1). This is seen in Figure 5.2 where, for example, on day 1, H_1 and H_6 are in roles R_1 and R_6, respectively, but on day 6, their roles are reversed. And since H_6 is above some or all of H_7, H_8, H_9, H_{10} on days 2–6, H_1 is never above any of them. In general, still using mod 10 subscript arithmetic, on day i, $H_i = R_1$ and $H_{i+5} = R_6$, but on day $i + 5$, their roles are reversed: $H_{i+5} = R_1$, $H_i = R_6$, also, since H_{i+5} is above H_{i+6}, H_{i+7}, H_{i+8}, and H_{i+9},

Day→	1	2	3	4	5	6	7	...	i	...	i+5
Role↓ R1	H1	H2	H3	H4	H5	H6	H7		Hi		Hi+5
R2	H2	H3	H4	H5	H6	H7	H8		Hi+1		Hi+6
R3	H3	H4	H5	H6	H7	H8	H9		Hi+2		Hi+7
R4	H4	H5	H6	H7	H8	H9	H10		Hi+3		Hi+8
R5	H5	H6	H7	H8	H9	H10	H1		Hi+4		Hi+9
R6	H6	H7	H8	H9	H10	H1	H2		Hi+5		Hi+10=Hi

Figure 5.2 The rotation of ten households through six roles. (Subscript arithmetic is mod 10.)

H_i is never above them. Similarly, H_{i+5} is never above H_{i+1}, H_{i+2}, H_{i+3}, and H_{i+4}.

After 1900, the number of households in the cooperatives decreased as a result of the overall decrease in the number of community households. Even with fewer households in each cooperative, cycling through the various roles would still insure equality of time and work contributions. The two criteria for the equality of status, however, could not be met without adjusting the number of roles. Using a bit of algebra, we can find a relationship linking n, the number of households in the cooperative, with r, the number of roles necessary to meet these criteria. Both status criteria are satisfied if $2r - 2 = n$; neither criterion is met if n is less than $2r - 2$; and only the stipulation of who should not be above whom holds if n is more than $2r - 2$.

The ideal Basque situation, where $r = 6$ and $n = 10$, clearly satisfies both status criteria and the relationship $2r - 2 = n$. And, while we do not know how the Basque arrived at their numbers, the Basques also knew that there could be at most five roles when there were eight households, four roles when there were six households, and three roles when there were four households. To accommodate odd numbers of households and the situations where there were more than the necessary minimum of households, they loosened the stipulation of the complete role reversal of woman of the house and house servant (that is, they used an n greater than, rather than equal to, $2r - 2$). The number of roles was reduced in about 1900, from six roles to five roles by deleting R_4, and then in about 1940, they were further reduced to four roles by deleting R_5. In the 1960s and 1970s, they were still further reduced by either reassorting the functions into three newly titled, but still hierarchically ranked roles, or by creating only two roles by combining into one the master ranks R_1 and R_2, and into another the servant ranks R_3 and R_6.

Thus, we see in the Basque concept of equality, a relation that is not static but is, rather, a dynamic process of interaction involving rotation, serial replacement, and alternation. An essential feature of the systemized interactions is that the participants know what is expected of them and what to expect from others. That is, the actors in the process move in synchronization, doing different things at different times, but together making up a whole. If one were to stop the process at an arbitrary time, there would be inequities in what has been contributed, what has been received, and who is superior to whom. But, just as a circle is enclosed by a never-ending line, the process of creating an equal–equal relation continues throughout the seasons and throughout the years.

4 Quite different from the Basque concern that there *not* be fixed status differences, the Tongans of Polynesia emphasize status, rank, and hierarchy. For the Tongans, knowing who is above whom is paramount; the hierarchy is defined by a formal system of *inequality* relations.

In some mathematical systems, as well as in some social systems, inequality relations play a significant role. Before seeing the expression of these relations among the Tongans, let us look at the concepts of inequality, order, and ranking more generally. In mathematics, inequality extends beyond the single relation "a is not equal to b," and includes comparatives, such as "a is greater than b" or "a is less than b." It is these comparatives that enable the imposition of order on a set of elements. In general, in common daily activities, we are probably far more concerned with numerical inequalities than with equalities. When I go to make a purchase and have, say, \$7.50, I do not say that I will buy the item if the cost *equals* \$7.50, but, rather, that I will purchase it if the cost is *less than or equal* to \$7.50. I drive through a road underpass only if its height is *greater than* the height of my vehicle. The algebra of inequalities for the real number system (that is, for the infinite set of numbers that can be expressed in decimal form) is different from the algebra of equalities. Here, however, we need only concern ourselves with a few basic ideas that link inequalities to order, and order to hierarchy and ranking.

The concept of order has already been encountered in earlier chapters. In Chapter 1, an *ordered pair* of numbers (a,b) was first used to identify the destiny spirits in the Caroline Islands (see Section 1.2). It was noted that in each pair which number was written *first* and which *second* was significant, as well as the values of the numbers. Earlier in this chapter (Section 4.3), the roles taken on by the men in the Basque shepherding group while on the mountain were said to be ranked. To reflect the ranking order, *we* introduced integer subscripts and identified the roles as R_1, R_2, ..., R_6.

Thus, the words *first* and *second*, and subscripts 1,2, ..., 6, have already been used in discussing order because the positive integers exemplify the concept. The positive integers are said to be *simply ordered* or *linearly ordered*. Using the relation *less than*, they satisfy an axiom called the *Law of Trichotomy*, which states that for any two integers, *a* and *b*, *one and only one* of the following hold:

(i) *a* is less than *b*, or

(ii) *b* is less than *a*, or

(iii) *a* is equal to *b*.

Also, for the positive integers, the less than relation has the transitivity property; that is, if *a* is less than *b*, and *b* is less than *c*, then *a* is less than *c*. It is these two—the trichotomy law and transitivity—that characterize order.

For the integers, we could, and often do, involve the relation *greater than* as well as *less than*, since the statements *b is greater than a* and *a is less than b* are equivalent. The trichotomy stipulation then is that *a* is less than, greater than, or equal to *b*. And, in addition to integers, the concept of order can be extended to other sets of items. For example, the points on a straight line are ordered; for any two points on the line, one is to the left, or the right, or coincident with the other. And, if point *a* is to the left of point *b*, and *b* is to the left of *c*, then *a* is to the left of *c*.

For the collection of letters we call the English alphabet, we stipulate an arbitrary order; namely, a, b, c, d, ..., z. We carefully teach that order along with the names and shapes of the letters. Hence, to users of this alphabet, what is meant by alphabetical order seems obvious. It is only when we encounter collections of letters from the alphabet of another culture, such as the collection α, δ, μ, ω, that we realize there is nothing obvious about alphabetical order. Military ranking systems are also learned, arbitrary orders. For example, consider the command structure of the U.S. Army. There are privates, corporals, sergeants, lieutenants, captains, majors, colonels, and generals. Just as the words are learned, so are their relative positions in the hierarchy. Since both alphabetical order and the military order are linear orders, both satisfy the trichotomy law and transitivity.

Clearly, not all collections are ordered. In mathematics, the elements contained in collections referred to as *sets* have no intrinsic order. For some sets, an arbitrary order or ordering relation is specified. As we noted, for the positive integers and for the real numbers, the ordering relation is *less than*. For some collections, no ordering relation is possible. One such case, for those who have encountered it, is the complex number system. Another instance is the dates in the Balinese wuku year on the dates in the Maya Calendar Round (both discussed in Chapter 3). Neither satisfy the trichotomy law—these dates cycle, and so, for any pair of dates, each date in the pair comes both before and after the other. As a last example, consider a competition between teams A, B, and C in which a ranking is to be based on who beats whom in games played between pairs of teams. Suppose two games are played in which A beats C, and then C beats B. This could only lead to partial ranking because there is no information about what the outcome would be if A played B. However, if A and B did play, with the outcome that B beat A, no ranking would be possible—although the trichotomy law is satisfied, transitivity is not. Alternatively, if A won, then both the trichotomy law and transitivity are satisfied, and the rank order is A, C, B.

With these ideas in hand, we turn to the ranking system of the Tongans. It is of particular interest because it involves the interrelation of several ranking principles.

5

The Tongan archipelago in the South Pacific (see Map 1 in Chapter 4) consists of about 150 islands, many of which are uninhabited. In all, the population of about 100,000 people inhabits a land area of about 700 sq. km (270 sq. miles). Tonga lies between Samoa and Fiji. Although culturally distinct, and with related but different languages, the three are linked in a number of social exchanges.

The Tongan concern for ranking is deeply embedded in the culture, as evidenced by their language and all of their interpersonal relationships. The words to use when speaking to someone; the attitude to have toward them; mutual expectations and responsibilities; ownership and the distribution of goods; roles in ceremonies; who has the say over matters of naming, marriage, and death rituals; all of these and more are determined by relative ranks. What makes the system particularly complex is that it contains three subsystems that are in force simultaneously. One subsystem is a formal ranking of those within a kin group; that is, of those related through descent. Quite separate is a ranking of named social groupings, which applies across the culture. And also for the culture as a whole, there is a ranking of those who govern. The latter two are easiest for us to state, although there is considerably more detail within each than is included here. In the governing ranking, there is a *king* who outranks the other *chiefs* who outrank the *ceremonial attendants* who outrank the *commoners.* Cutting across the ranks of chiefs and commoners are ranked title-holding groups. Not everyone is part of, or interested in, this subsystem. These named groups have political and ceremonial functions and are of importance primarily to those involved in seeking or exercising governing powers. For a Tongan, setting and context determine which of the subsystems dominate in any particular circumstance. At a funeral, for example, kin ranking is of paramount importance, while at a kava drinking ceremony, it is the ranked titles that prevail.

Here, we look primarily at the ranking subsystem that applies to kin. We will, however, return to the governing ranking to discuss a problem that arises from the interplay of these two subsystems.

The kin ranking relies on three characteristics: relative generation; relative age; and gender. The fundamental relations are:

1. Within the same generation, sisters outrank brothers.
2. Within the same generation and same sex, elders outrank those who are younger.
3. A father and his kinsmen outrank his child who outranks the mother and her kinsmen.

To symbolize the *outranks* relation, we borrow the *greater than* symbol. In this shorthand, the first relation above can be written as Si > Br (sister outranks brother); the second as E > Y (elder outranks younger); and, for the last, Fa > C > Mo (father outranks child who outranks mother). As with *greater than*, the *outranks* relation is not reflexive and not symmetric, but, where it applies, it is transitive; that is, if A outranks B and B outranks C, then A outranks C (if A > B and B > C, then A > C).

When two characteristics are involved, we form a composite representation containing both. That is, for example, reading from right to left, SiE would read *elder sister*. Then, assuming four siblings (two sisters and two brothers), relations within a single generation of siblings can be summarized as:

$$\text{SiE} > \text{SiY} > \text{BrE} > \text{BrY}. \qquad (1)$$

These composites can be extended to include three characteristics. For example, FaSiE would read *elder sister of father* so that

$$\text{FaSiE} > \text{FaSiY} > \text{Fa and Mo} > \text{MoBrE} > \text{MoBrY}. \qquad (2)$$

Thus, limiting ourselves to the perspective of the children, combining the generations,

$$\text{FaSiE} > \text{FaSiY} > \text{Fa} > \text{SiE} > \text{SiY} > \text{BrE} > \text{BrY} > \text{Mo} > \text{MBrE} > \text{MBrY}. \qquad (3)$$

Additional siblings in both generations could be added easily, but their gender and relative ages would have to be known or assumed.

In some traditional Western families, when daughters learned the roles of their mothers, they became mothers' helpers in the household. They learned, for example, to cook for their fathers and brothers, as their mothers did. In Tonga, however, this is definitely not the case. A female must not prepare food for her brother; it is he who cooks for her. In general, brothers learn to exhibit extreme respect for their sisters. Notice in (3) that, with respect to a young or grown child, it is the father's eldest sister who is the dominant person. It is she who has the say in issues of naming and of marriage. And, at a funeral, she dresses distinctively, distributes the funeral goods, and has first choice of the goods that are distributed. For a woman, her lowest status is as a wife and mother, while her highest status is as a sister and an aunt.

To view the relationship of a person to those in the generation below, the person's gender must be considered. For a woman, we know that she is below her children, but as an aunt, she outranks her nieces and nephews. If, however, the individual is a male, he is, as we already know, above his own children. His brothers' children are classified with his own, and so he is above them as well. His sisters' children, however, are above their mother who, as we already know, are above their brothers. And among the sisters' children, it is the eldest daughter who outranks the others. Thus, a male has particular obligations to his sisters' children, and his eldest sister's eldest daughter is of most importance.

As for relative ranking among what we call cousins, recall that children have the same ranking relation to parents' kinsmen as to parents. Hence, one is outranked by cousins on one's father's side, while outranking cousins on one's mother's side. And within the cousins, one's father's eldest sister's eldest daughter ranks the highest, and one's mother's youngest brother's youngest son ranks the lowest.

The rankings are of importance, for the most part, in regard to one's own generation and to the generation above and the generation below. Generations, however, are interwoven so that, for example, a woman may be a daughter, sister, and mother, while her mother, in turn, is a daughter, sister, and mother, and similarly, her daughter links to yet another generation. The obligations, behaviors, and responsibilities based on kin ranking primarily come into play in the realm of personal goods and services and decisions and ceremonials surrounding life passages. Inheritance of houses, lands, and titles, for example, does not fall within this subsystem.

Although largely there is no conflict between the subsystems, a problem arises for the king. The problem is that in the kin ranking subsystem, his sisters and his sisters' children outrank him and his children. This can interfere with the king's dignity and honor and can create the potential of a challenge to his or his childrens' power. The Tongan solution is that a king's sister is not permitted to marry, or a spouse must be found for her such that her children will fall *outside* the Tongan system. At the start of this section, we noted that Fiji, Tonga, and Samoa were linked in some social exchanges. One important linkage is that when a high-ranking female in Samoa must be "married out," she marries a Tongan, while the sister of a Tongan king marries a Fijian. Her spouse is chosen from a chiefly line among Fijians, but his children are outsiders to the Tongan power structure. In effect, for purposes of finding a solution to the dilemma, the Tongan system is embedded in a larger system.

6 For another system of social organization, quite different from that of the Basque and Tongans, we move to Africa and the Gada system of the Borana of Ethiopia. The system is extensive: it encompasses the philosophical, ritual, and political life of the group. The system particularly attracts our attention because of its formality, the interplay within it of linear and cyclic components, and its clear articulation by Borana historians. Furthermore, in Gada, we see how a conceptual model is superimposed on reality. In contrast to the previous sections in which the particular relations of equality and inequality were highlighted, here, our emphasis is on viewing the overall system as a complex of relations with interconnections among them.

The Borana are a branch of the Oromo people who live primarily in Ethopia. In all, there are about 15 million Oromo who speak related dialects of the Galliñña language. Some branches other than the Borana, are the Macha, the Guji, the Afran Kallo, and the Wollo. Most of these groups follow some version of the Gada system. Our discussion deals only with the practices of the Borana who live in Sidamo province in Ethiopia and south into Kenya (see Map 5.1). They largely depend on cattle herding, and, since the land is arid, there is considerable attention devoted to wells and water.

Basic to the Gada system are consecutive grades, through which,

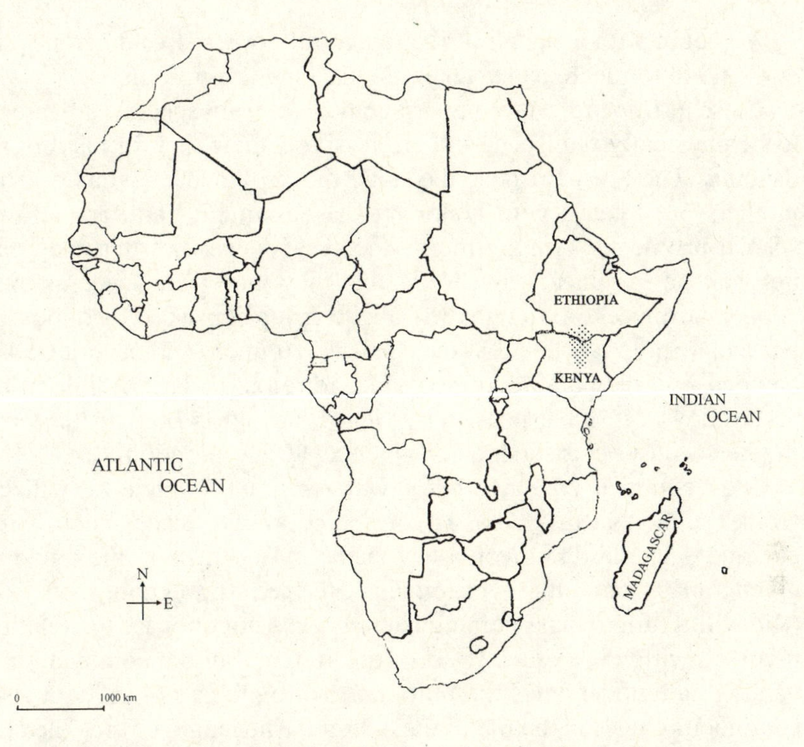

Map 5.1 Africa. (The shaded region is inhabited by the Borana.)

theoretically, all males pass, and named classes, which refer to those who are in the grades. The distinction between *grades* and *classes* is significant and crucial to our understanding of the system. An analogy, used solely to reinforce the distinction between grades and classes, is that all United States college students, theoretically, pass through the consecutive *grades*, freshman, sophomore, junior, senior. While so doing, and even later, they are identified as, say, *Class* of '56 or *Class* of '02. As with the Gada system, although the grades are somewhat related to age, not everyone in each grade is the same age. Also similar is that not everyone in the same class enters the college at the same time or in the same grade, and not all class members participate in all class activities. The analogy, of course, is quite limited. The systems differ considerably in their extent and in their cultural significance, as well as in what characterizes the grades and how they are related, the ceremonials that take place as a class passes from grade to grade, how classes are determined, their relation to each other, and how they are named.

We begin with a sketch of the temporally ordered Gada grades that correspond to idealized stages in the passage through life.

Those in *Dabelle*, a preliminary grade, are given special care, are not to be physically punished, and are to spend their time in playing and dancing. They remain near the huts, the province of women. Their mothers are treated with honor by the community and are allowed special privileges. Among the mothers' privileges are that they need not wait on the lines at the wells, and they may wear twice as many copper hairpieces. Although the Dabelle group consists only of boys, of particular importance is that their hair is left uncut and decorated with cowries, and they are addressed and referred to as girls. While in this grade, boys can have no sisters. Hence, any girls born to the boys' mothers' must be abandoned or given away for adoption.

Upon entrance into the *Junior Gamme* grade (which we will call Grade 1), their birth as sons is considered to take place. Their hair is cut, and each child is given a new name. Also, those in the grade are given a class name that will remain with them throughout their lives. (We will return to class naming later because not only are the relations involved within it a crucial part of the system, but the combination of naming and grades links the finite linear life process of individuals to continuous cycles in the life of the culture.) The length of time the class spends in Grade 1 is 8 years. During this time, grade members stay around the cattle enclosures, the province of men, and may help herd small livestock. Now, it is permissible for sisters to be born. Toward the end of the 8-year period, the grade members, under careful supervision, may begin to participate in cattle raids, hunts, and war parties.

The next grade, *Senior Gamme* (Grade 2), also lasts for 8 years. During that time, the grade members visit each other, participate in rowdy behavior, harass those in the grade below them, and begin to act as warriors. Toward the end of the grade, a most important selection takes place; leaders of the group are chosen. There are six councillors, six deputy councillors, and several ritual positions. The selection is of major consequence because, at the transition to *Cusa* (Grade 3), the leadership roles become formalized and will continue beyond the grade.

Upon entering Grade 3, the chosen leaders change their dwelling places to live in close proximity to each other, and where the leaders live becomes the capitol of the class. No decision or ritual relating to the class will take place without the leaders. From the six councillors, a top councillor is chosen, and the class becomes identified with his name. During the 8 years of Grade 3, those in the grade assist their fathers in pastoral chores, begin to take a role in rituals and perform appropriate sacrifices, engage in big game hunts, and, although under supervision of those in the grade above, are predominant in warfare and cattle raids. Their outfits become those of adults, and much time is devoted to finding mistresses and looking for wives. Marriage, however, cannot take place before the beginning of Grade 4 (*Raba*).

At the beginning of Grade 4, the class is married. In this ceremony, *as in all others*, the leaders must participate. All others in the class participate if they can and if they wish to—they may defer their marriages to a later time. (As with numerous rituals and ceremonies, the leaders represent the class; what they have done, the class is considered to have done.) During the 8 years of Grade 4, the members of the grade, married or not, are not permitted to have children. This is strictly enforced. Those in the grade serve as senior warriors and, in general, engage in wild and aggressive behavior.

In Grade 5 (*Dori* or *Senior Raba*), the group's attention turns to family life and to learning about governance and matters in the public sphere. It is now legitimate and appropriate to have sons. Daughters, however, are forbidden and are abandoned or given away for adoption. It is important to note that participation in prescribed rituals and ceremonies, which had its beginnings in Grade 3, continues to be of importance throughout the grades.

Grade 5 lasts only for 5 years, with the next 3 years an initial part of Grade 6 which lasts, in all, for 11 years. Together, then, Grades 5 and Grades 6 last for 16 years with Grade 6 being the *Gada* grade during which those in the grade bear major responsibility for the entire community. During the initial 3 years of grade 6, the members participate in a big-game hunt and in numerous ceremonies. There is a gradual handover of responsibilities to them from those who have just passed out of the grade into the next.

At the end of this 3-year period, there is a formal ceremony marking the end of the transition and the beginning of new leadership. The ceremony requires that the incoming and outgoing classes camp next to each other for several weeks. At a prescribed time, everyone, except for the two class leaders, must remain indoors while the two leaders exchange milk and blessings, and the scepter of authority is formally handed over. The sacrifice of a bull marks the conclusion of the transition. Soon after the takeover, all member of the incoming group undergo circumcision and earpiercing, and then must remain indoors for about a month, eat special foods, make special offerings, and participate in a ceremony marked by the release of snakes. It is during this time that their sons formally enter Grade 1. *Thus, all sons are considered to be born when the fathers are 40 years old.* But "40 years old" is *not* the father's biological age; it is the father's age as measured from *his* formal birth into the Gada system.

The primary role of those now in power is to maintain "The Peace of Boran", which deals with the relationship between the nation and God, as well as to resolve any disputes that may arise between descent groups, clans, classes, or camps. The conflicts may be of any type—ritual, political, moral, legal, or economic. The leaders also must see to the planning and execution of communal projects, such as the excavation of a new well. In all, it is a heavy burden and one that is willingly passed on at the end of the 8 years. These 8 years in the Gada grade are of such significance that the historical period will be identified by the name of the class in the grade and the name of its top councillor. (The significance of the grade is underscored by the fact that both the overall system and the grade are referred to by the word *Gada*.)

Grades 7, 8, and 9 are, together, the *Yuba* grades lasting in all for 24 years. Those in these grades are considered wise and responsible advisors, as well as ritual experts. Grade 10 (*Gadamoji*), also lasting 8 years, is one in which the members do no herding, engage in no public affairs, use no weapons, control their tempers, and moderate all their behaviors. Then comes *Jarsa* (Grade 11), a sacred state that lasts indefinitely.

In all, there are ten grades lasting 80 years and a preliminary grade and a final grade that are variable in length and, in effect, outside of the system. A man's passage into Jarsa comes at the same time as his sons' assumption of power and his grandsons' official birth into the system. As we can see from looking at Figure 5.3, the conceptual model superimposes symmetry and clarity of form on the stages of life. By having a

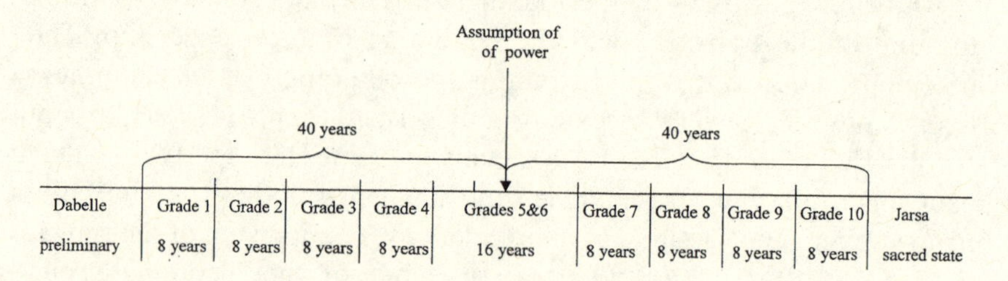

Figure 5.3 Gada grades.

specific beginning and end, and clear demarcation points, the grades are well defined and ordered in time.

It should be evident when reading the description of the grades that not all life passages of every individual can take place in conformity to the ideal. Relationships between men and women other than spouses are not uncommon or unacceptable. They simply exist alongside the formal model. Children do get born from these relationships and at times that are considered improper. Most marriages are not at the designated time for the class. Not only may some of the men in the class be too young, but since all brothers are in the same class, their fathers may have insufficient means to provide the required bride wealth for all of them simultaneously. And, since polygyny is accepted, with men often having senior and junior wives, sons may actually be born well after their class has passed through several grades. As a result, men may miss participating in the ceremonies, rituals, activities, and responsibilities that go with those grades. However, since it is not biological age but class that determines one's role, young boys and older men may share on an equal footing in discussions, decision making, and responsibilities. No matter what the biological reality, the crucial assumption of the system is unchanged—*all sons are born and enter a named class 40 years after the birth of their fathers.*

With this outline of the grades in hand, we now turn to the class names used by the Borana historians, how they are assigned, and how they are conceptualized. The interrelationships of the names link the past to the present and the present to the future.

As was already noted, at the time that those in Grade 6 assume power, their sons are born into the system and given a class name. The name is one of only *seven* class names. The class names occur in a fixed order and recur in a cycle. For ease of identification, we will call them $(c_1, c_2, c_3, c_4, c_5, c_6, c_7)$. The subscripts reflect the fixed order, and the parentheses are a conventional mathematical way of indicating that the enclosed items cycle. Hence, which name is written first in the parentheses is arbitrary—$(c_4, c_5, c_6, c_7, c_1, c_2, c_3)$, for example, represents the same cycle of seven names, as does $(c_6, c_7, c_1, c_2, c_3, c_4, c_5)$. A diagram of the cycle is in Figure 5.4. The class name given to a son is *always* five names behind the class name of his father. Thus, if a father is in class c_i, $(i = 1, 2, \ldots, 7)$, his son is in c_{i-5}, where, because of the seven-name cycle, the subscript arithmetic is mod 7. For example, if a father is in c_7, his son is in c_2,

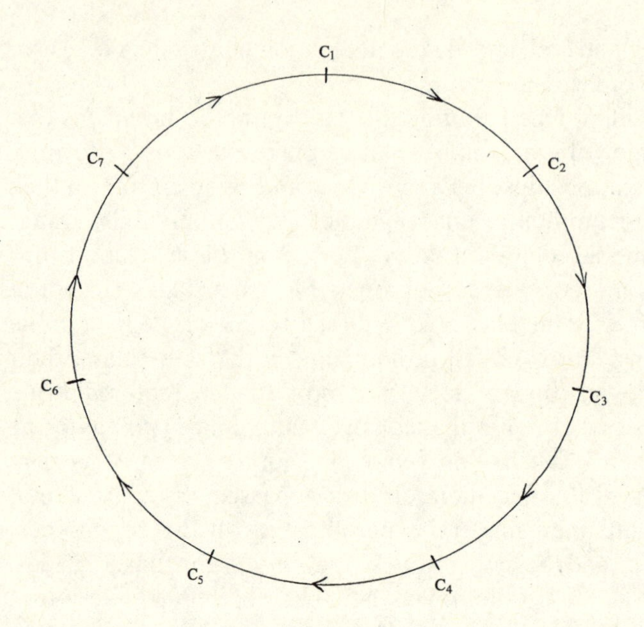

Figure 5.4 The class name cycle. (→ leads to the name that follows.)

and a father in c_3 has a son in $c_{3-5(\text{mod } 7)} = c_5$. The relation between fathers and sons can be encapsulated by:

$$S(c_i)=c_{i-5 \text{ (mod 7)}} \text{ for } i= 1,\ 2,\ \ldots,\ 7 \tag{4}$$

where $S(c_i)$ is the son of c_i. Similarly, a man in c_i can be related to his grandson (the son of the son of c_i) by

$$S(S(c_i))=S(c_{i-5}) = c_{i-10 \text{ (mod 7)}}$$

or, in general, using an exponent to indicate the number of generations that pass, that is, the number of times S, "the son of," is applied:

$$S^n(c_i) = c_{i-5n \text{ (mod 7)}} \text{ for } n = 1,\ 2,\ldots;\ i = 1,\ 2,\ \ldots 7. \tag{5}$$

With the foregoing relationship and a bit of modular algebra, it can be determined when, and if, some descendant of a man in c_i returns to

c_i. There will be a return to c_i if

$$c_i = c_{i-5n \text{ (mod 7)}},$$

which occurs for any solutions of

$$i = i - 5n(\text{mod } 7).$$

That is, it occurs after n generations for all n satisfying

$$5n = 0(\text{mod } 7). \tag{6}$$

In modular algebra, cancellation must be approached with caution. If, say, $2 = 6(\text{mod } 4)$, by cancelling a factor of 2 on both sides, one might erroneously conclude that $1 = 3(\text{mod } 4)$. Or, by simply cancelling a 3 on both sides of $3n = 9(\text{mod } 6)$; the solutions $n = 1(\text{mod } 6)$ and $n = 5(\text{mod } 6)$ could be missed. Cancellation of k, however, causes no difficulty in $ka = kb(\text{mod } m)$ where k and m are *relatively prime*, that is, where they have no common factor. Also, for k and m relatively prime, solving $kn = kb(\text{mod } m)$ for n by cancellation is possible; it yields an integral solution such that any other solutions are equal to it mod m.

In the case being considered here [seeking the solution to (6)], 5 and 7 are, indeed, relatively prime. Hence, it can be concluded that the solution is $n = 0(\text{mod } 7)$; that is, $n = 7$ or any multiple of 7. Therefore, starting with any c_i, the father-to-son descent line does return to c_i after seven generations, and every seventh generation thereafter. It, too, is a cycle of length seven containing all seven class names. The order in which the names follow each other in this cycle, however, differs from the class name cycle: it is $(c_3, c_5, c_7, c_2, c_4, c_6, c_1)$. (See Figure 5.5.)

In mathematical terminology, the father-to-son descent cycle is a *permutation* of the class name cycle. A permutation, in general, is a one-to-one transformation of a finite set of elements onto itself. We could, for example, transform the name cycle into the father-to-son descent cycle by replacing each c_i by $c_{2i+1(\text{mod } 7)}$. There are other ways (which the reader might wish to explore), but what is important is that both cycles are of length 7, and, in each, each of the seven names appears once and only once (compare Figures 5.4 and 5.5).

Rather than focusing on the fact that the class name of the sons is five names behind that of the father, we can, instead, focus on the fathers and say that the class name of the father is five names ahead of his sons. That is,

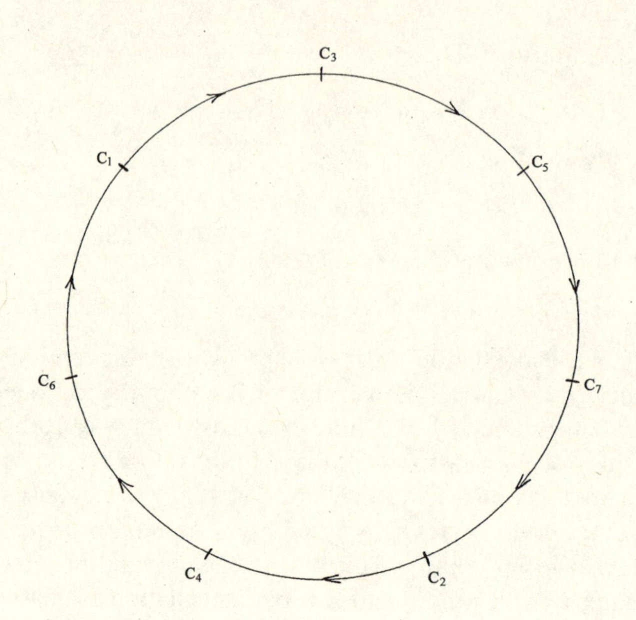

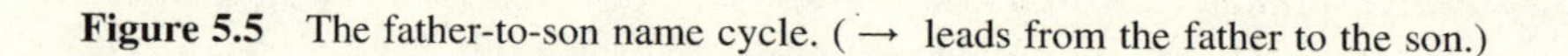
Figure 5.5 The father-to-son name cycle. (→ leads from the father to the son.)

$$F(c_i) = c_{i=i+5(\text{mod } 7)} \text{ for } i = 1,\ 2,\ \ldots,\ 7 \qquad (7)$$

where $F(c_i)$ is the father of c_i. In the Borana system, because class names rather than individual names are involved, and all sons have the same class name, "the son of" and "the father of" are inverse relations. This means that beginning with some c_i and consecutively applying the relations $S(c_i)$ defined in (4) and $F(c_i)$ defined in (7) always returns to the original c_i:

$$F[S(c_i)] = F[c_i - 5] = c_i \quad \text{and} \quad S[F(c_i)] = S(c_i + 5) = c_i.$$

In other words, the class name of the father of the son of a man in c_i is c_i, and the class name of the son of the father of a man in c_i is c_i. In general, for individual names, the father and son relations are not inverses. Let us see why with individual names the situation differs. Say, for example, that the father's name is Jacob, his father's name is Isaac, and his son's name is Joseph. Then, provided that Isaac has only the one son, Jacob:

$$F[S(\text{Jacob})] = F(\text{Joseph}) = \text{Jacob}; \text{ and } S[F(\text{Jacob})] = S(\text{Isaac}) = \text{Jacob}.$$

However, if Isaac has two sons, say Jacob and Esau, then there is ambiguity:

$$F[S(\text{Jacob})] = F(\text{Joseph}) = \text{Jacob}; \text{ but } S[F(\text{Jacob})] = S(\text{Isaac}) = \text{Jacob or Esau}.$$

In the Borana system, *all* of a father's sons have the *same* class name, and so this problem does not arise.

Were the focus on $F(c_i)$ rather than on its inverse $S(c_i)$, the same cycle of names would result, but it would move in the opposite direction: (c_5, c_3, c_1, c_6, c_4, c_2, c_7). In Figure 5.5, the arrows move in the positive direction of time, that is, from fathers to sons, moving against the arrows—back in time—would lead from the sons to the fathers. From either perspective, although each individual is in only one class throughout his life, he is linked to all classes through his ancestors and descendants.

At any specific time, because each group retains its class name as it moves through the grades, consecutive class names are held by consecutive grades. There are, however, more grades than class names, and so all names are present, but some will be present more than once. Let us look, for example, at a time point when c_7 is the name of the class holding power in Grade 6. The names of the classes in the respective grades would be as shown in Figure 5.6.

The class name of the fathers of c_7 is c_5, but clearly, the fathers are not those named c_5 at this time. The present c_5 are in a lower grade (Grade 4). And, c_2 appears twice, but we know that the c_2 sons of c_7 are those in Grade 1, not those in Grade 8. There are, then, two distinct sequences present: the c_2, c_3, c_4, c_5, c_6 in grades 1–5 and their respective fathers c_7, c_1, c_2, c_3, c_4 in grades 6–10. In general, *at any given time*, there are two related *five*-name sequences present. The five-name sequences play a significant role in the Borana conceptualization of the grade system as the culture moves through time.

class name	c_2	c_3	c_4	c_5	c_6	c_7	c_1	c_2	c_3	c_4
grade	1	2	3	4	5	6	7	8	9	10

Figure 5.6 Classes in grades at a specific time.

c_2	c_3	c_4	c_5	c_6
c_7	c_1	c_2	c_3	c_4
c_5	c_6	c_7	c_1	c_2
c_3	c_4	c_5	c_6	c_7
c_1	c_2	c_3	c_4	c_5
c_6	c_7	c_1	c_2	c_3
c_4	c_5	c_6	c_7	c_1
c_2	c_3	c_4	c_5	c_6
.	.	.	.	.
.	.	.	.	.
.	.	.	.	.

Figure 5.7 Consecutive class names in groupings of five.

In Figure 5.7, using the same time point as is represented in Figure 5.6, beginning with the Grade 1 class c_2, we write the consecutive class names in a row, but move to a new row after each set of five names. (The consecutive-class names, as we know, are in a seven-name cycle.) The first seven rows are all different, but the eighth row duplicates the first. We stop writing after the eighth row, but could continue indefinitely. (Recall that in Chapter 3, Section 1, there is a discussion of the length of a supracycle resulting from combining cycles of different lengths.) Notice that as the rows are formed, we are moving back through time. The second row contains the class names of the fathers of the first row, the third row contains the fathers of the second row, and so on. Looking down each of the five columns, we see the names in the order of the son-to-father cycle, beginning with the names of the classes in grades 1–5, respectively. The way that this is written on Figure 5.7 is ours, but the interplay of groupings of five with cycles of seven forming a supracycle of 35 belongs to the Borana.

Each of these five son-to-father lines is called a *gogessa* by the Borana. Usually, men can state the class names of their fathers to a few generations. Borana historians, however, using this conceptual framework, can locate events in the past as distant as fifty or more 8-year periods. They associate an event with the class name and leader name of those in the Gada grade at the time of the event. This is of special importance because it is believed that events of specific earlier time periods have profound implications for current times. Just as the same configuration of names in our 7×5 array reappears after 35 periods have passed, so do the influences of the events and outcomes of that earlier time. The history of 280 years ago is not just an interesting story: it is crucial cultural knowledge. *Dacci* is the Borana word for the influence of history on current happenings. That influence is transmitted through father-to-son lines, but operates, in particular, in a cycle of 35 periods.

Thus, for example, a Borana historian tells that the time period identified with class c_6 led by Morowwa Abbayye "returns upon" Jaldessa Liban (the leader of class c_6 in the Gada grade 35 periods later). He continues:

> Morowwa's gada was a time of peace and plenty. Morowwa dug a well called El Morowwa. All Borana drank from that well. After Morowwa died, the well disappeared, and the Galantu have been short of water ever since. It is this [gada] of plenty that returned upon Jaldessa Liban. As you can see today, both man and cattle are fertile and water is plentiful. Jaldessa Liban has found the well of his ancestor and Borana has, once again, begun to drink to satiation. The gada of Morowwa is very good, and we do not expect anything to go wrong in the present gada.

The relations that are reflected in Figures 5.4, 5.5, and 5.7, as stated by the same Borana historian (and translated into English by an African social scientist), are presented below. (Included in brackets are some of the statements recast into the terminology and symbolism that we have been using.)

> The *makabasa* are seven: *moggisa, sabbaka, libasa, darara, mardida, fullasa, makula.* The present gada is fullasa, before that was makula, before that was moggisa, before that sabbaka, libasa, darara, and mardida. [The class names are seven: $c_1, c_2, c_3, c_4, c_5, c_6, c_7$. Presently in Grade 6 is c_6, before that was c_7, before that c_1, before that c_2, c_3, c_4, c_5.]
>
> 1. The makabasa pass from father to son. Mardida is born to libasa, libasa is born to moggisa, moggisa is born to fullasa, fullasa is born to darara, darara is born to sabbaka, sabbaka is born to makula, makula is born to mardida. [The class names pass from father to son. $c_5 = S(c_3)$, $c_3 = S(c_1)$, $c_1 = S(c_6)$, $c_6 = S(c_4)$, $c_4 = S(c_2)$, $c_2 = S(c_7)$, $c_7 = S(c_5)$.]

2. The Makabasa return to the same gogessa after seven fathers. [The same class name returns to the same son-to-father line after seven fathers.]
3. Before it returns to the gogessa, the makabasa goes to the other four gogessa. [There are five son-to-father lines. A name goes to the other four lines before returning to its original position.]
4. The makabasa never goes to the gogessa of your *walanna*, and it never comes from the gogessa of your walanna.
5. When the makabasa returns, dacci also returns. It returns from gogessa walanna and gogessa *kadaddu*.
This is how the makabasa give birth to each other.

The words that are new are *walanna* and *kadaddu*. According to the Borana, there is competition and rivalry between adjacent classes, whereas alternate classes are allies or friends. The same characterization applies to adjacent and alternate generations and alternate gogessa. *Walanna* means rivals and *kadaddu* means allies. Rule 4, then, says that consecutive appearances of a class name are always in alternate—rather than adjacent—son-to-father lines.

In order to assure ourselves that this alternation is *always* true, we observe that for thirty-five consecutive class names arranged in a seven-row by five-column array (beginning with n_1 in row 1, column 1 and calling them n_j where $j = 1, 2, \ldots 35$), the column in which a name appears is $j(\text{mod } 5)$. The use of mod 5 results from moving back to column 1 at the end of each row of 5. Since the class-name cycle is of length seven, the next appearance of name n_j is $n_{j+7.}$ The column in which it appears is $(j + 7)(\text{mod } 5)$, which equals $j(\text{mod } 5) + 7(\text{mod } 5)$, or $j(\text{mod } 5) + 2$. That is, it is two columns beyond $j(\text{mod } 5)$. Similarly, the previous appearance of n_j is two columns before $j(\text{mod } 5)$. Look, for example, at c_7 in Figure 5.7. It first occurs as the sixth name (n_6), and so is in column 6(mod 5), that is in column 1. Its next appearance is as n_{13}, and so it is in column 13(mod 5), that is, column 3, and then in column 5, column 2, column 4, and then again in column 1. We not only see that any particular name does, indeed, go to and come from alternate columns, but also see that it returns to its original column after seven fathers (rows). And, as is stated in rule 5, when the name returns after a complete 35-period cycle, it brings with it the influence of that earlier time. It secondarily brings the influences of the times of the earlier appearances of the name during the cycle; those appearances were in both adjacent and alternate columns.

The Gada system clearly serves to integrate Borana culture. Its structure ties together the realms of the social, the political, the economic, and the philosophical. Our interest in it, however, is as a system of well-defined relations and relations between relations. The temporally ordered grades, the cyclic class names, and, even more so, the stipulation that each son is exactly 40 years behind his father in the system underlie the relations. The 40-year spread unites the grades and class names: 40 years spans five grades and places the sons five class names behind their fathers. There are seven class names, and they cycle (see Figure 5.4). Combining the *seven*-name cycle with a *five*-name displacement yields the son/father seven-generation name cycle (see Figure 5.5). Also, because sons are five class names behind their fathers, there are five distinct son/father lines present at any given time. Since the son/father cycle is of length seven, and the names within it are five consecutive names apart, the cycle spans *35* consecutive names. Thus, there is the thirty-five-name cycle of Figure 5.7. The fact that the numbers five and seven are relatively prime is particularly significant to the conformation of the system.

7 Systems of social organization, like mathematical systems, are human creations. Although there is no culture without some form of social organization, the systems vary considerably in their scope, their formality, and, above all, their configurations. Just the few we discussed should give some sense of how conceptually diverse they are, and that they are, indeed, arbitrary complexes of relations and relations between relations.

For the Basque, the concept of circle integrates the system of cooperation. The circle is a shared abstraction; each households *knows* which neighbor is their first neighbor and which is their second neighbor. The participation of a shepherding group's members and their roles are determined by a cycle. And going up and down the mountain is conceived of as a rotation. For the Tongans, however, hierarchical ranking dominates. And, for the Borana, there is division into classes and passage through consecutive grades, plus the arbitrary assumption that sons are born 40 years after their fathers. Within the systems, the various relations are built upon the unifying conceptions, reinforcing and operationalizing them.

Each of the systems also has, as mathematical systems must, clearly defined elements. Among the Basque, the members of a shepherding group, for example, are fixed through time. That is, each group has a specified set of members. As for neighbors in the circle, that identity remains with a house, regardless of who is occupying it at any given time. Also, as was noted, in the Gada system, grade and class membership is quite well defined, and, where there may be individual differences, the elected leaders stand for the group.

Among mathematicians and those who contemplate mathematics, the question is sometimes raised as to whether mathematics is discovered or invented. My own view is that both occur: mathematical systems are created (a more appropriate term than *invented*), but then, resulting from the relations within them, further relations are discovered to exist. Social systems are surely the creations of the people within whose culture they are found. But, no doubt, as time goes on, additional logical implications of the relations become apparent. Among the Basque, the complete role reversal during the first phase of the shepherding (that is, whoever is R_6 when H_i is R_1, is R_1 when H_i is R_6), may, for example, have been found to result from the other relations. For the Tongans, the problem inherent in the system for the king was probably unintentional. And, in the Gada system, consecutive appearances of a particular class name in alternate son-to-father lines might be an example of a discovered property rather than one that is part of the primary conception of the system. These, of course, are only surmises. But, in any system, no matter how well described, there are always ramifications of the relations within it that are yet to be explored and yet to be discovered.

NOTES

1. Useful discussions of mathematical relations are found in the college texts *Foundations and Fundamental Concepts of Mathematics*, H. Eves, 3rd edition, PWS-Kent, Boston, MA, 1990, pp. 132–136; *The Anatomy of Mathematics*, R.B. Kershner and L.R. Wilcox, 2nd edition, Ronald Press, New York, 1974, pp. 50–60; and *Rethinking Mathematical Concepts*, R.F. Wheeler, Ellis Horwood, Chichester, UK, 1981, pp. 11–23. The last is directed toward prospective teachers. A more general and scholarly book is C.E. Rickart's *Structuralism and Structures: A Mathematical Perspective,* World Scientific, London, 1995. In a section entitled "The Basic Definitions" (pp. 17–20), he discusses relations as the essential ingredients of structures. His definition of a structure (p. 17) is "any set of *objects* (also called *elements*) along with certain *relations* among those objects." As he also notes, "the definition is considerably more subtle than its simple form might indicate." As for a system, it (p. 19) "is a collection of interrelated objects along with all of the *potential* structures that might be identified within it"; that is, it "may be perceived in more than one way as having structure, depending on which properties are singled out for attention." These definitions and observations are very important to the understanding of our definition of mathematical ideas, which includes *those ideas involving number, logic, spatial configuration, and more significant, their combination or organization into systems and structures.* And, when we, as mathematicians, see some similarity in social systems, we are singling out for attention only very limited aspects of those systems.

2. Discussions of equality and the equivalence relation are on pp. 34–36, 284–287 in the Kershner and Wilcox text noted in the Section 1 notes above. In the Eves and Wheeler texts, they are included in the sections already cited.

Sociopolitical discussions of equality that were influential in Euro-American culture are, for example, *Nicomachean Ethics*, Book V, Aristotle, fourth century BCE; Jean Jacques Rousseau's "A Discourse on the Origin of Inequality" (1754) and "The Social Contract" (1762); and John Stuart Mill's "On Liberty" (1859). The phrases quoted as common American-English usage appear under the synonyms for *same* on p. 1289 of *Webster*'s *New World Dictionary of the American Language,* College Edition, World Publishing Co., New York, 1966.

3. My discussion here of the Basque and their ideas is drawn from my article "What does equality mean?—The Basque view," *Humanistic Mathematics Network Journal* #18, Nov. 1998, pp. 22–27. Both the article and this discussion are derived from my reading of *A Circle of Mountains: A Basque Shepherding Community*, Sandra Ott, Clarendon Press, Oxford, 1981. Of particular relevance are pp. vii–viii, 1–41, 63–81, 103–106, 129–170, and 213–217. The few phrases directly quoted are from p. vii. Other mathematical ideas of the Basque are being studied extensively by Rosyln M. Frank, University of Iowa. See, for example, "The Geometry of Pastoral Stone Octagons: The Basque Sarobe," Rosyln M. Frank and J.D. Patrick, pp. 77–91 in *Archeoastronomy in the 1990s*, Clive L.N. Ruggles, ed., Loughborough Group D Publications, London, 1993, or "An essay on European ethnomathematics: The coordinates of the septuagesimal cognitive framework in the Atlantic facade," Rosyln M. Frank, 78 pp., ms., 1995. Also, a special counting technique among the Basque living in California is described in "Counting sheep in Basque," Frank P. Arawjo, *Anthropological Linguistics*, 17 (1975) 139–145.

4. A discussion of order relations can be found in chapter 15 of *The Anatomy of Mathematics* cited in the Section 1 notes above, as well as in most abstract algebra texts.

5. For additional understanding of the Tonga ranking system and its manifestation in the culture, the writings of Adrienne L. Kaeppler are recommended: "Rank in Tonga," *Ethnology*, 10 (1971) 174–193; "Exchange patterns in goods and spouses: Fiji, Tonga and Samoa," *Mankind,*11 (1978) 246–252; and "*Me*'a *faka*'*eiki*: Tongan funerals in a changing society;" pp. 174–202 in *The Changing Pacific*, Niel Gunson, ed., Oxford University Press, Oxford, 1978. Also see Garth Rogers' "'The father's sister is black': A consideration of female rank and power in Tonga," *Journal of the Polynesian Society*, 86 (1977) 157–182.

In chapter 2 of *Structural Models in Anthropology*, Per Hage and Frank Harary, Cambridge University Press, Cambridge, 1983, after introducing terminology and concepts from graph theory, the Tonga kin subsystem is used as an example of a transitive tournament and acyclic digraph (pp. 80–83). The book is intended to enable social scientists to use graph theoretic ideas. For teachers and students of graph theory, it can broaden their view of the subject's applicability. In their 1991 book, *Exchange in Oceania*, Clarendon Press, Oxford, the same authors discuss the Tonga example in more detail as part of their discussion of the logic of relations (pp. 263–268). The authors are an anthropologist and a mathematician, respectively. Their writings are recommended.

6. My discussion of the Gada system of the Borana is primarily based on *Gada: Three Approaches to the Study of African Society*, Asmarom Legesse, The Free Press, New York, 1973; *Authority and Change: A Study of the Kallu Institution Among the Macha Galla of Ethiopia*, Karl E. Knutsson, Ethnologiska Studier, vol. 29, Ethnografiska Museet, Göteborg, Sweden, 1967, pp. 160–169; and "Boran age-sets and generation sets: Gada, a puzzle or a maze?" P.T.W. Baxter, pp. 151–182 in *Age, Generation and Time: Some Features of East African Age Organizations*, P.T.W. Baxter and Uri Almagor, eds., St. Martin's Press, New York, 1978. Although they are about the Gada system as used by a neighboring Oromo group, two additional useful references are *Guji Oromo Culture in Southern Ethiopia*, Joseph Van de Loo, Dietrich Reimer Verlag, Berlin, 1991, and "The Guji: Gada as a ritual system," John Hinnant, pp. 207–243 in the book edited by Baxter and Almagor cited above.

The book by Legesse is of particular interest because it is an extensive description and analysis and is written by an African social scientist trained in anthropology. He includes a chapter entitled "An essay in protest anthropology" in which he critiques the ethnocentrism found in the foundations of anthropology and in anthropological writings. He believes, however, that, with care, many of the tools and insights that were developed can be used to further the understanding of the African heritage. (Included in Legesse's book is a discussion of the Borana calendar. I did not find this part of his discussion clear or convincing. A critique of his discussion is "The Borana calendar: Some observations," C.L.N. Ruggles, *Archeoastronomy*, 17 (1987) S35–S54.) The statements from the Borana historian, Arero Rammata, quoted here were obtained and translated by Legesse. They appear on pp. 198–199 and p. 192, respectively, in his book and are quoted with his permission. (He notes that the specific class names used by the expert historians differ from those used by others. Also, the bracketed word in the first quotation is my substitution for the word *fullasa*, which is not introduced here until later.)

Modular algebra, more formally known as the algebra of congruences, is also discussed in Chapter 3, Section 3. As included in the notes for that section, for basic discussions, see *Invitation to Number Theory*, Oystein Ore, New Mathematical Library, MAA reprint of 1967 original, and *Elements of Number Theory*, I.A. Barnett, Prindle, Weber, & Schmidt, Boston, MA, 1969. For a more extended introduction to permutations, cycles, and congruences, *Introduction to Mathematical Structures and Proofs*, Larry J. Gerstein, Springer, Sudbury, MA, 1996, pp. 211–243, 281–321 is especially recommended.

The father-to-son cycle also has the properties of a cyclic group of order 7. Moreover, in group theoretical terminology, it is an Abelian group and, because 7 is prime, it is a simple Abelian group. For an introduction to groups, see *Groups and Their Graphs*, Israel Grossman and Wilhelm Magnus, New Mathematical Library, 1964, reprinted by Mathematical Association of America, Washington, DC. Also, in *Ethnomathematics: A Multicultural View of Mathematical Ideas*,

Marcia Ascher, Chapman & Hall/CRC, New York, 1994, the concept of a group is presented in detail, and the Walpiri kin system is shown to be a group. In contrast to the simple Abelian cyclic group of order 7 noted here, the Walpiri system is a dihedral group of order 8. (*The Genesis of the Abstract Group Concept*, Hans Wussing, MIT Press, Cambridge, MA, 1984, about the history of the group concept within Western mathematics, is highly recommended. It helps us to understand that *our* view of what constitutes mathematics changes through time, and, in particular, how the emphasis on structure evolved.)

CHAPTER 6

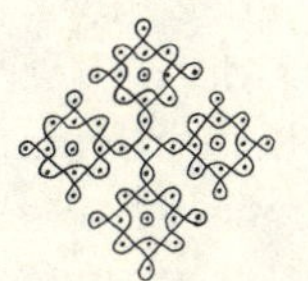

Figures on the Threshold

Each morning, women in Tamil Nadu, in southern India, place designs on the thresholds of their homes. The designs, known as *kolam*, are created by using rice flour held in the hand and slowly trickled in a thin stream from between the index and middle finger. More than simply a folk art, the kolam tradition is closely tied to, and expressive of, the values, rituals, and philosophy of the people of Tamil Nadu.

One of the first written references to kolam is in the sixteenth century, although the origin of the tradition may be far earlier:

> Once there was a king, Vallālmakārājan, who ruled in Aruṇai... He was an excellent king, truthful, benevolent, always praising Śiva's feet; he cared for all lives as if they were his own; he had no desire for others' wealth, and he regarded all women other than his wives as if they were his sisters. With such a fine king, it is no wonder that the kingdom flourished: the tiger and the cow drank from the same watering place, Brahmins chanted the Veda, women decorated the streets with *kolams*, rain fell on schedule, and the hungry were fed.

Even if not always adhering to all the details of the traditional practice, the creation of threshold decorations continues today with the kolam designs well known to those who live in cities, as well as those who live in rural areas, and to college students and professionals, as well as those who have had less schooling.

The kolam tradition is mathematically interesting for several reasons. The complex and intricate figures are intriguing in and of themselves, but, what is more, in many cases, their creation involves the transformation and superimposition of basic subunits, and there are families of kolam whose members can be derived from each other in patterned ways. Replicating the richness of these figures and their growth patterns became a challenge to computer scientists who were creating picture languages and were studying formal language theory. In addition to their importance to the people of Tamil Nadu, the creation of kolam designs have become part of the computer science literature, serving as examples for some types of languages and as the inspiration for additional types of languages. Thus, the kolam provide an exemplar of the way that mathematical ideas in a traditional setting can reach beyond their own cultural boundaries to enrich and contribute to scholarly interests. At the same time, of course, the attention from this new perspective has deepened and enriched our understanding of the kolam. Here, we focus on both the kolam tradition and on its linkage to picture languages.

1 There are several traditional art forms spread throughout India that are similar to kolam. There are, for example, the closely related *rangoli* or *rangavalli* in the Deccan region just north of Tamil Nadu, *aniyal* in Kerala to its west, and *alpana* in Bengal. They share some characteristics: they are practiced by women and originally utilized rice and rice flour. There are, however, some differences in traditional configurations, methods of construction, and specific cultural meanings. Although they are probably historically related, we confine ourselves to the kolam tradition of the people of Tamil Nadu in the southeastern region of India. Prior to India's independence from the British in 1956, this region was part of what was called the Madras Presidency. The region then became the state of Madras until 1969, when its name was changed to Tamil Nadu.

For mathematicians familiar with the work and story of Srinivasa Ramanujan (1887–1920), the region has additional significance because it is where he was born, raised, educated, and died. With the exception of his stay in Cambridge from 1914 to 1919, Ramanujan lived surrounded by, and immersed in, Tamil Nadu culture. He is known to have quoted proverbs and allegories from folk tales of Tamil Nadu while discussing mathematical problems.

The official language of the state is Tamil, and, as contrasted to other cultures we have discussed, there is a written script and an extensive Tamil literature. Although the writings go back to the third or fourth century BCE, the earliest references to kolam that have been found in Tamil literature occur much later. These, however, are only passing references and do not describe kolam in detail. Kolam, itself, is part of the people's oral tradition, which exists side by side with its textual tradition.

The figures are referred to as *kolam*, but in Tamil, the word is not limited to them but is broader in meaning and usage. The meanings include *beauty*, *gracefulness*, *form*, *shape*, and *appropriate dress*, and are linked to the concept of order, which is viewed as a significant aspect of beauty. A house without kolam is called *mūli,* that is, a woman without nose-rings, ear-rings, and, if married, without the red dot on her forehead.

The women and girls of a traditional Tamil Nadu household begin their day by sweeping the floor and the area in front of the house, sprinkling the threshold area with a solution made with cow dung, and then decorating the threshold area with kolam. The women are familiar with numerous designs—some for daily use, some for special occasions, and some for particular rituals and holidays. Girls are taught the designs and techniques by their mothers. It is an important part of a girl's training, as kolam skills are considered a mark of grace and a demonstration of dexterity, mental discipline, inner harmony, and an ability to concentrate. The elaborate threshold decorations, which are built with the basic designs and techniques, are aesthetic and creative expressions of the women who "place" them. Although deemed less traditional and less creative, today some additional sources of designs are used, namely commercial magazines and even stencils.

The materials used to create the designs have significance. Cow dung, because of its germicidal properties, is to cleanse and disinfect the floors. The use of rice powder at the threshold shows kindness to inferior insects because it is food for ants. At the same time, it keeps the ants from entering the house. (Most recently, powdered limestone or some other coarse white powder, lacking in significance, is used.)

The decorations at the threshold serve both to welcome guests on auspicious occasions and, on other occasions, as a protective screen, averting misfortunes and illness and keeping evil spirits from the house. Philosophically, the threshold is a boundary, but a permeable one, symbolic of the boundary between the inner world of the mind and emotion and the outer world of the landscape and action, or as the place of passage from the sacred to the mundane. It is also identified with transition points in the cosmic cycle, related both to the lives of individuals and to the seasonal round. Thus, beyond their aesthetic value, the decorations on the threshold are chosen to appropriately mark life-passage events, rituals, or as the prelude to worship of particular deities. There are no kolam in front of a house at the time of a severe illness or death of a family member.

The threshold decorations are particularly elaborate during the *ponkal* festival. The month before the festival is considered especially unlucky, in part because it is the coldest and dampest month with infectious diseases common. The 3-day festival ends that month, but it also marks the winter solstice, the Tamil New Year, the transition from darkness to light, and the transition from an unlucky month to a lucky one. It is a time of thanksgiving, visiting, and rejoicing.

Some of the kolam represent objects; others represent animals, birds, or various flowers or vines. Where the names of the designs are available, we will include them. The designs, individually and as groups, are rich in symbolic meaning. The multiple meanings are not always as familiar to contemporary practioners, nor as accessible to us, as are the designs themselves. Two symbolic meanings, however, have particular significance for our interest in the construction processes of the kolam.

In one type of kolam—*pulli kolam*—a grid of dots is first placed on the ground. The dots are referred to as *pulli*. The configuration of the grid is an important guide to placing the rest of the figure. Some kolam are constructed by drawing lines connecting the pulli; for others, the lines go around the pulli. Many of the kolam for which the lines go around the pulli are made of a single continuous closed curve, and others are made with a few continuous closed curves. These kolam are related by interpreters to the continuous, never-ending cosmic cycle of birth–fertility–death, and with the concepts of totality, perfection, and eternity. To some, the pulli represent the source, or raw potential.

The importance of the pulli as defining elements of the kolam, and the kolam as a familiar and distinctive element of Tamil Nadu culture, is highlighted in the opening sequence of a late 1980s political film made in India, in Tamil Nadu. In the sequence, the screenwriter, who, for several years, was chief minister of the state, addresses the viewers and concludes his comments by referring to the kolam and their framework of pulli; he notes that just as those who place the kolam are constrained by the dots, he must work within the limits of the law "to expose the hidden but powerful forces that are exploiting the nation."

There are, in addition, other kolam, placed without dot grids, which are also made up of one or a few continuous closed curves. These, too, are associated with continuity and never-ending cycles.

2 Before we embark on the picture languages, how and why they became involved with the kolam tradition, and the specific kolam that attracted study by computer scientists, let us look at a few kolam designs in order to get some sense of their style and configurations.

Figure 6.1 shows a group of kolam that are all different, yet share similarities of component parts. In fact, close inspection shows that,

(a)

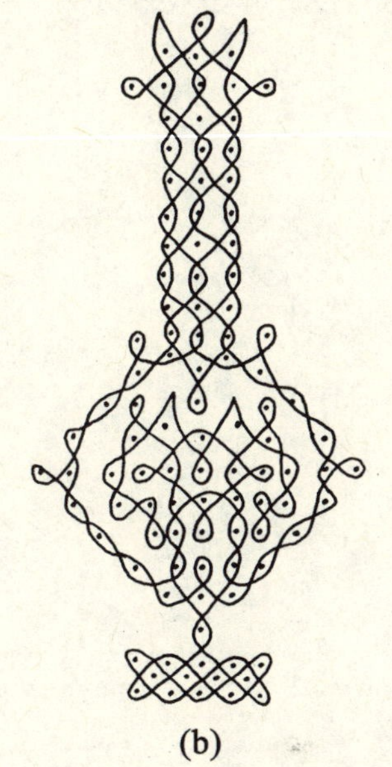

(b)

(c)

(d)

Figure 6.1 Some kolam. (a) Sandal Cup; (b) Rosewater Sprinkler; (c) Hanging Lamp; (d) Nose Jewel.

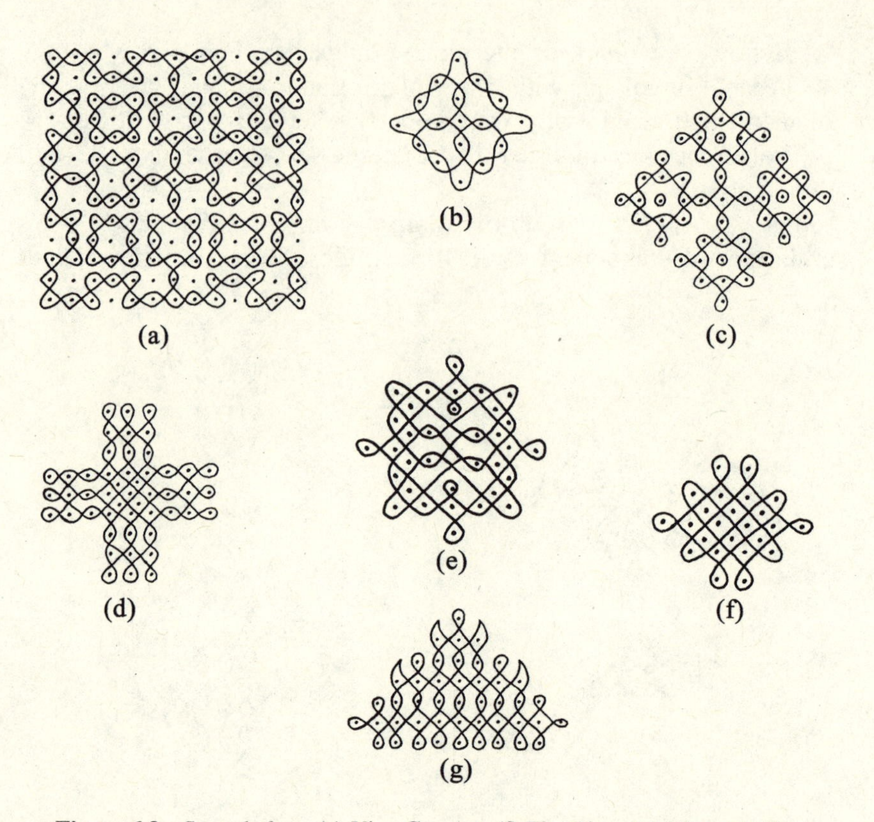

Figure 6.2 Some kolam. (a) Vine Creeper; (f) The Ring; (g) Mountain Top.

with slight modification, the "Nose Jewel" (Figure 6.1d) is part of the "Hanging Lamp" (Figure 6.1c). All the designs in Figure 6.1 also are symmetric with respect to a central vertical line. Figure 6.2 is another group showing some similarities while being different. These, too, all have vertical symmetry, but, with the exception of Figure 6.2g, they all also have symmetry across a central horizontal line. Also, each of the first four, that is Figures 6.2a–6.2d, also show 90° rotational symmetry around a central point. Notice, also, that Figure 6.2a contains almost all of Figure 6.2c within it. The kolam in Figure 6.3 are quite different from those in Figures 6.1 and 6.2, and different from each other. They do, however, all show rotational symmetry around a center point, but the angles of rotation differ. For Figures 6.3a and 6.3c, there is 180° rotational symmetry; for Figure 6.3b, it is 90°; and for Figure 6.3d, it is 45°. As we go on, we will show several kolam that are larger and more elaborate; we will also discuss them in greater detail.

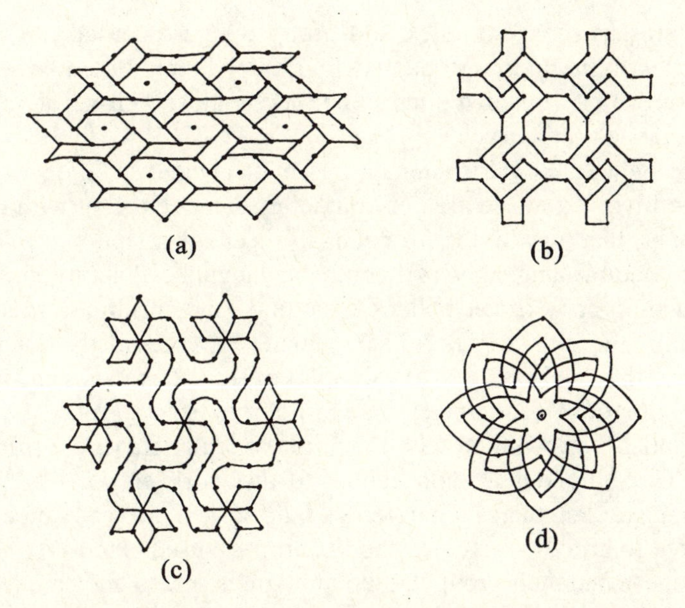

Figure 6.3 Some kolam. (a) Mango Leaves; (b) Asanapalakai; (c) Parijatha Creeper; (d) Lotus of the Heart.

3 Gift Siromoney of Madras Christian College in Tamil Nadu created the bridge between kolam and computer science. Throughout his life, Siromoney combined his academic specialty with his deep interest in the culture, history, and environment of Tamil Nadu. A commemorative volume, published soon after his death in 1988, contains a selected bibliography of about 100 publications. These include, but are by no means limited to, computer recognition of Tamil and Brahmi script, statistical studies of South Indian sculptures and of Indus texts, and computer methods for dating Tamil inscriptions. Siromoney also pursued an investigation of the kolam tradition, studying its history, as well as the manner in which contemporary women constructed and remembered the designs.

In one study, Siromoney compared the perception of kolam forms, in terms of complexity and of similarities and differences, for two groups of women—those familiar with kolam and those unfamiliar with them. Because the designs were so well known to women in Tamil Nadu, the "knowers" who participated in the study were 19 Indian female college students and the "nonknowers" were from out of the area, namely 14 American undergraduates from North Carolina who were attending a Term-in-India program. In general, the American undergraduates saw the figures as more complex and grouped figures based on size, as contrasted to the Indian students who grouped them based on whether they were made up of a single continuous closed curve or multiple continuous closed curves.

With his wife, Rani Siromoney, a computer scientist, Gift Siromoney became involved in the area of computer science dealing with picture languages, that is, with the formal analysis and description of pictures. Akin to natural languages and computer languages, picture languages are made up of restricted sets of basic units and specific, formal rules for putting the units together. The kolam designs provided a rich supply of figures that could be used as examples of some languages and require the creation of new languages. The Siromoneys and a group of computer scientists who worked with them, in particular Kamala Krithivasan and K.G. Subramanian, contributed to the work on what are called Lindenmayer languages (L-systems), but focused primarily on extending these to array L-systems. The kolam were used extensively by the Siromoneys and their group, but kolam figures spread and can be found as examples in the work of others.

The use of picture languages to generate kolam brings to the fore the structures within specific kolam or within families of kolam. The languages are, essentially, concise statements of algorithms, or formulas, for generating kolam. They make us more aware that the kolam are careful constructions with definable growth patterns. It is important to keep in mind, however, that the computer scientists are concerned with analyzing and generating the figures, and are not necessarily replicating the techniques or thought processes used by Tamil Nadu women.

4 The use of formal languages became of broad interest with the work of linguist Noam Chomsky in the late 1950s. His concern was the linguistic analysis of natural languages. He wished to separate grammatical sequences of words in a language from ungrammatical sequences by finding a set of rules that would generate all grammatical sequences, but none of the ungrammatical sequences. Chomsky's work incorporated the idea of using rewriting rules for strings of symbols. Computer scientists became especially interested in formal languages when they realized that they, too, were using the same ideas in establishing formal definitions of newly created programming languages. These works, and the works of others, eventually led to the picture languages of interest to us. An important contribution along the way was the work of Aristid Lindenmayer, a biologist interested in modeling plant growth.

Let us look first at the rudiments of a simple formal language that is not pictorial, but produces only strings of symbols. Then, we will see how these strings of symbols can be translated into pictures.

There is, for each language, a fixed set of symbols. The set of symbols is referred to by different names; we will call it the *alphabet*. There is also a string of symbols that one starts with; a starting string is generally referred to as the *axiom*. Each language has a set of rules for creating new strings of symbols from previous ones. These we call *rewriting rules*. And, there are what we will call *outcomes*, which are the strings of symbols that result from applying the rewriting rules.

Here is an example of a string language with an alphabet containing just the three symbols A, B, C. (These symbols, then, are the only ones that we can use.) Our starting string—the axiom—will be ABB. The rewriting rules we will use are A → BC, B → A, and C → C. This says that to create a new string from a previous string, replace each A by BC, each B by A, and each C by C.

1. Start: ABB
2. Result of applying rewriting rule once: BCAA
3. Result of applying rewriting rules to the outcome BCAA: ACBCBC
4. Result of applying rewriting rules again: BCCACAC
5. And so on.

Here the letters are only symbols; they have no arithmetic or other meaning. Creating a string of symbols by placing them adjacent to each other is referred to as forming their *concatenation*. Here that simply means that one is followed by the next. Notice that in each step, the three rules are applied simultaneously (commonly referred to as *in parallel*), not sequentially. That is, in each step, the B and C newly introduced by applying the rule A → BC remain unmodified until a later step. This is characteristic of an L-language (Lindenmayer language) as contrasted with some other languages. The language in our example is also of the type called *context-free* languages because each symbol is dealt with individually without reference to the neighboring symbols. And, it is a *deterministic* rewriting system because there is only a single possible rewriting rule for each of the symbols.

Here is another example of a deterministic, context-free, Lindenmayer string language. For this example, we will use the alphabet F, +, –, with starting string F − F, and rewriting rules F → F + F, + → +, − → –.

Start: F − F
Outcome 1. [F + F] − [F + F]
Outcome 2. [(F + F) + (F + F)] − [(F + F) + (F + F)]
Outcome 3. F + F + F + F + F + F + F + F − F + F + F + F + F + F + F + F
Outcome 4. F + F + F + F + F + F + F + F + F + F + F + F + F + F + F + F − F + F + F + F + F + F + F + F + F + F + F + F + F + F + F + F
⋮
etc.

The parentheses and brackets used above are *not* symbols in the alphabet or in the outcomes; they are only included here to highlight the substitutions. Again, it is important to keep in mind that, so far, the symbols F, +, and – have no arithmetic or other meaning.

Our next important step is to move from a string of symbols to a picture and, hence, to picture languages. A way to do this, developed by Przemyslaw Prusinkiewicz, is to interpret the symbols as "turtle" commands. Turtle graphics was originally an innovation introduced to engage children in the creative use of computers. In it, the turtle is thought of as sitting on a piece of paper, facing in some starting direction. The turtle can carry out a limited set of commands, such as move forward while drawing a line, move forward without drawing a line, turn left, and turn right. (The turtle's tail is dirty. Hence, lowering his tail draws a line as he moves, while raising his tail leaves no line.) Specifically, here are the commands our turtle understands and the symbols that convey those commands:

F: move forward by step length *s* while drawing a line;
f: move forward by step length *s* without drawing a line;
+: turn left (counterclockwise) through an angle of *d* degrees;
−: turn right (clockwise) through an angle of *d* degrees.

For each drawing, the start direction, step length, and turn angle must be specified; the step length and turn angle remain the same throughout the drawing. When following a string of commands, each turtle move begins in the place and direction that the last one ended.

Let us assume that our turtle begins by facing to the right, that the step length is one unit, and that the turn angle is 90°. Then, the turtle interpretation of the axiom F − F in our previous string language example is: move forward one unit while drawing a line, turn right 90°, move

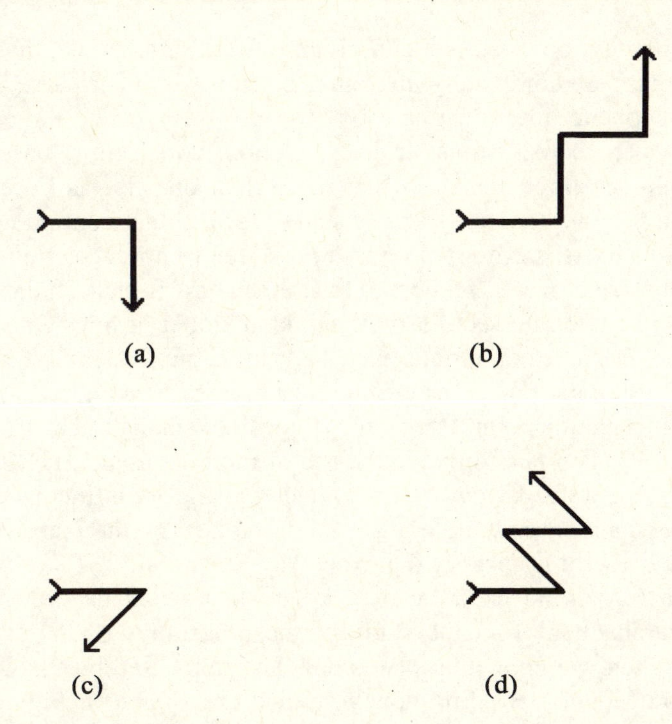

Figure 6.4 Turtle drawings. (a) F − F ($d = 90°$); (b) F + F − F + F ($d = 90°$); (c) F − F ($d = 135°$); (d) F + F − F + F ($d = 135°$).

forward one unit while drawing a line. The pictorial result is shown in Figure 6.4a. Figure 6.4b shows the pictorial result of the first outcome, F + F − F + F. If, instead of using a turn angle of 90°, we change the specified angle to 135°, the pictorial results would be as in Figures 6.4c and 6.4d.

5 With this brief introduction to picture languages, we return to the kolam and focus, in particular, on some known as *kambi* (wire) kolam. The name reflects the fact that these kolam are created using a single, continuous, closed curve. The curve may or may not intersect itself. In a sense, a kambi can be thought of as made of a never-ending line.

Before proceeding with the kambi kolam, however, we emphasize that they constitute a special category of kolam; that is, there are many other kolam that are made up of several closed curves or several lines rather than by a single, continuous line which returns to where it began. For some kolam, we can see by simply looking at them that more than a single line *must* be used. Notice in Figure 6.2a, for example, that parts of the kolam are completely disconnected from the other parts, and so several, separate lines must be used.

Even where there is no visible disconnection, we can know, based on a theorem from graph theory, when more than one line is necessary. According to that theorem, simply stated, for a single continuous line that returns to its starting point to be possible, each intersection in the figure must have an even number of lines emanating from it. (If the figure includes just two intersections with an odd number of lines emanating from them, a single continuous line can be used, but it does not return to its starting point.) Look, for example, at Figures 6.3a–6.3c. Each has several intersection points from which three lines emanate. Thus, each of these three kolam *must* involve the use of more than one line. But for those that theoretically can be drawn using a single continuous closed curve, we cannot know, unless it is so designated by the maker of the kolam, whether it is actually done that way.

Figure 6.5a is an example of a kolam that *could* be drawn as a single kambi, but, from the study of the procedures used by Tamil Nadu women, we *know* that it was not. The steps used by the women whose procedures Gift Siromoney studied are shown in Figure 6.5. After laying out a five-by-five grid of pulli, they proceeded to draw a closed curve (Figure 6.5b), and then repeated the same curve three times, but each time they rotated the curve through 90° (Figures 6.5c–6.5e). The kolam was completed by drawing a frame, another closed curve, around the interlocking curves—again, see Figure 6.5a. Hence, for this kolam, the 90° rotational symmetry that we see in the finished drawing is more than simply an externally imposed concept based on our viewing of the completed figure; the rotational symmetry is the characteristic of the figure motivating the construction procedure.

Figure 6.6 shows a family of kolam named “Anklets of Krishna.” Each of these *is* a kambi kolam. A *family* is a set of curves that are different, but related by some common characteristic. As you look at the curves, you can see that the larger curves are made up of several copies of the smaller curves. Perhaps you can even see that there is a pattern to the way the subpatterns of each are linked. These patterned repetitions are what intrigued the computer scientists. For this family, they sought a picture language whose outcomes would be the different members of the Anklets of Krishna family, and no others. Although the languages developed do not necessarily replicate the drawing

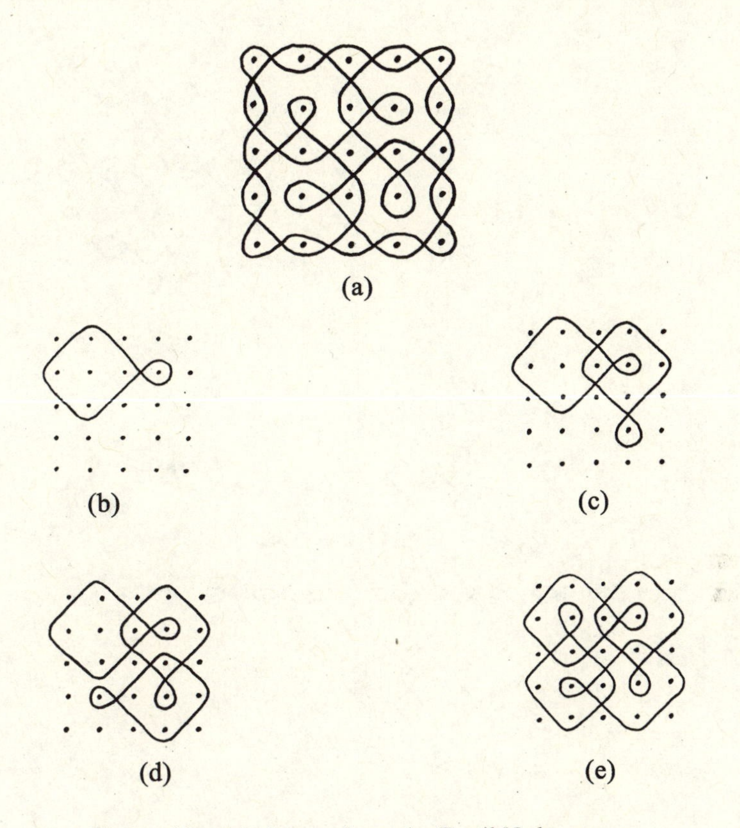

Figure 6.5 A kolam as drawn by Tamil Nadu women.

procedures used by the Tamil Nadu women, they do, none the less, give us insight into the inner structure of the figures.

Clearly, because of the definition of turtle moves, the pictures drawn by the turtle are always angular and made up of linear segments. Most of the kolam, however, like the Anklets of Krishna, are smooth curves, and so, some adaptations or other modes of drawing were used. Some computer scientists used the turtle moves and produced angular versions of the kolam. Another approach was to produce the angular versions and then smooth them using auxiliary techniques. Based on their study of the fundamental drawing units used by the Tamil women, the Madras group, instead, defined idealized "kolam moves", which result in smooth curve segments. The turtle, as conceived of in turtle geometry, cannot carry out these kolam moves. A creature that leaves a sinuous trail and fits Tamil mythology is a snake. The full complement of kolam moves and the symbols associated with them are shown in

(a)

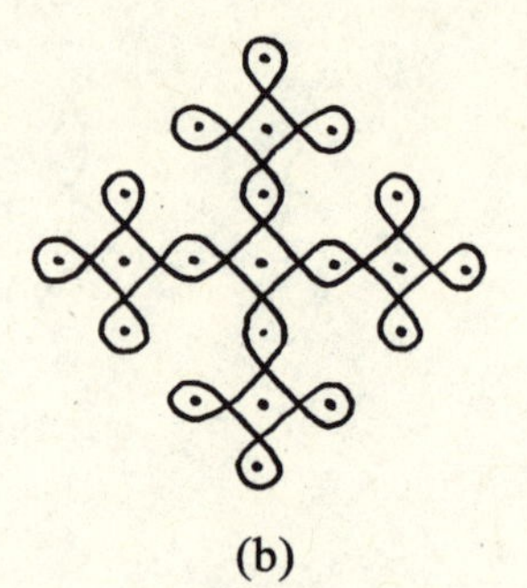

(b)

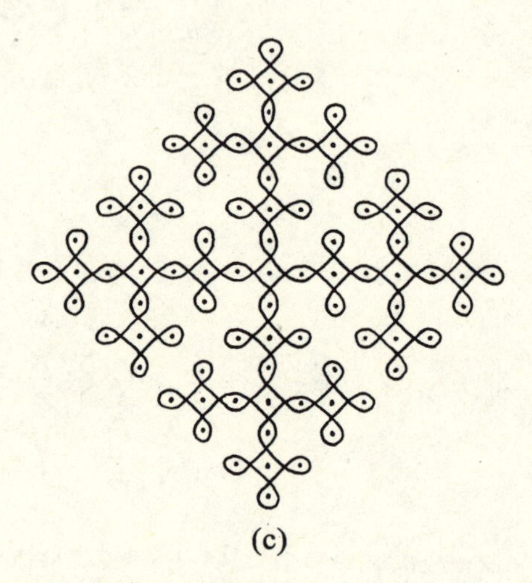

(c)

Figure 6.6 Anklets of Krishna. (The pulli are not produced by the string language. They are, however, produced by the array language described later.) (a) Axiom; (b) Stage 1; (c) Stage 2.

Figure 6.7. They are F (move forward one unit), R_1 (move forward while making a half-turn right), R_2 (move forward and make a u-turn to the right), R_3 (more forward while making a full loop to the right), L_1 (move forward while making a half-turn left), L_2 (move forward and make a u-turn to the left), and L_3 (move forward while making a full loop to the left). As with the turtle commands, a starting direction and unit length always must be specified. For the kolam move drawings and turtle move drawings that we use as illustrations, a convenient unit length will be selected to fit the page and make the details visible. That is, the unit length remains the same within *each* drawing, but may vary from drawing to drawing.

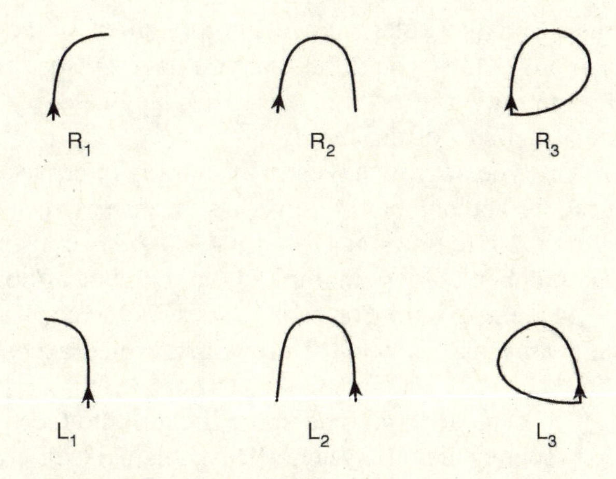

Figure 6.7 Kolam moves.

For the Anklets of Krishna, the only kolam moves needed are F, R_1, and R_3. The axiom for the language producing the family is R_1FR_3 FR_3 FR_3 FR_1, and the rewriting rules are $F \rightarrow F$, $R_1 \rightarrow R_1FR_3FR_1$ and $R_3 \rightarrow R_1FR_3FR_3FR_3FR_1$. Figure 6.6a is the pictorial representation of the axiom with the starting point and starting direction marked. Figure 6.6b is the pictorial representation of the outcome resulting from applying the rewriting rules just once; that is,

$$(R_1FR_3FR_1)F(R_1FR_3FR_3FR_3FR_1)$$
$$F(R_1FR_3FR_3FR_3FR_1)F(R_1FR_3FR_3FR_3FR_1)F(R_1FR_3FR_1).$$

Here, again, the parentheses are only to assist in identifying the substitutions.

This language concentrates on the figure and does not include the pulli, nor does it precede the figure by placing the pulli. (The pulli, however, are included in Figures 6.6a–6.6c as we will return to use them again to illustrate another approach.) Describing the pictorial representation of the axiom (Figure 6.6a) as a flowerlet with four petals, the pictorial effect of applying the rewriting rules is to replace each of the four petals by a four-petaled flowerlet (Figure 6.6b). In the next stage, created by applying the rewriting rules to the outcome above, each flowerlet is replaced by a set of four flowerlets. Thus, with successive rewritings, these grow from one to four to sixteen flowerlets. If the rewriting rules are applied a third time, the next member of the Anklets of Krishna family would have sixty-four flowerlets. You are invited to continue the process, using the rewriting rules and the kolam move interpretation of the resulting strings of symbols. Or, if you think that you have grasped the pattern expressed by the rewriting rules of the string language, try to directly draw the next member of the family.

The type of growth seen in this family of curves is called *exponential* growth. The number of flowerlets in successive stages grows from 1 to 4 to (4×4) to $(4 \times 4 \times 4)$ to $(4 \times 4 \times 4 \times 4)$, or, using exponents to show the number of fours multiplied, the number of flowerlets are 1, 4^1, 4^2, 4^3, 4^4, and so on. For each successive stage, it is the exponent that increases, and, hence, the growth is termed *exponential.*

6 Let us look at one more kolam produced by a deterministic context-free L-system. This kolam is different in kind, as it is one that is placed without pulli. Furthermore, although it is a kambi kolam, it is one that does not intersect itself. Named "the Snake," this kolam particularly attracts our attention because of its relationship to mathematical objects called *fractals*, and, what is more, to a special category of curves within them.

Figure 6.8 shows one member of the Snake family. When discussing the pictorial language that produces it, we will use turtle moves and,

Figure 6.8 The Snake kolam.

hence, produce an angular version of the kolam. We do this for simplicity and, also, to follow the path used in the computer science literature. There, after producing the angular version, they went on to apply a formal smoothing technique. We, instead, stop with the angular version, and leave the reader to imagine the smoothing.

For the turtle drawings of the Snake language outcomes, the specified angle is 45°, and the turtle begins by facing to the right. For convenience of writing and visualization, we add to the basic symbols F, +, and −, another symbol, A. A is always to be interpreted by the turtle as the sequence of commands F + F + F − − F − − F + F + F + F. The pictorial representation of A is shown in Figure 6.9a. In all, the L-system is:

axiom: A − − F − − A − − F
rewriting rule: A → A + F + A − − F − − A + F + A
first outcome: (A + F + A − − F − − A + F + A) − − F − −
(A + F + A − − F − − A + F + A) − − F
⋮
etc.

The turtle representation of the axiom is shown in Figure 6.9b, and the representation of the first outcome is in Figure 6.9c. In Figure 6.9d, the second outcome is shown. Focusing on the pictorial representation of A, and the way replications of it are joined together, can help to see how the rewriting rule gives rise to the visual results.

From the drawings, we can see that the growth is exponential, as was the case with the Anklets of Krishna. Each of the arms in the four-armed cross in the axiom's depiction is replaced by a four-armed cross. Then, each of the crosses within that figure is replaced with a set of four crosses just like itself. Thus, it grows from one four-armed cross (Figure 6.9b) to a cross made up of four four-armed crosses (Figure 6.9c) to another cross made up, in turn, of four of these (Figure 6.9d), and so on. Figure 6.8 is, then, the smoothed version of the third outcome, namely, a cross made up of subcrosses, containing, in all, 4^3 or 64 of the beginning four-armed crosses. Both the Anklets of Krishna and this example of exponential growth involve powers of 4, but that need not be the case. The growth could just as well be related to, say 3 or 5, as with 1, 3, 9, 27, … or 1, 5, 25, 125, … .

Theoretically, we could continue indefinitely to generate new figures by successively applying the rewriting rule to the symbol string just produced. In the resultant figures, at each stage, there would be four

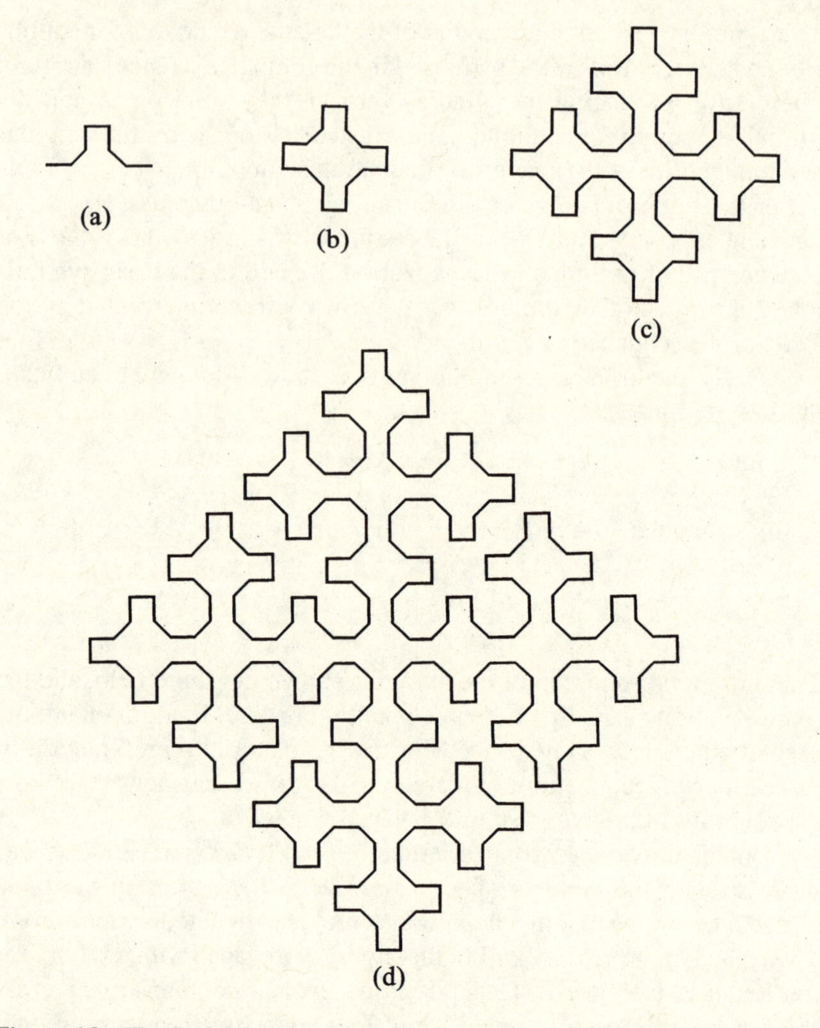

Figure 6.9 The Snake (angular version). (a) A; (b) Axiom; (c) First outcome; (d) Second outcome.

replications of the earlier stage combined in the same patterned way. Thus, we see that the overall process contains a recurring component, and the different resulting curves are expressions of the different number of times that component is repeated. The mathematical term for it is a *recursive* process; that is, it is one in which the result of a process is put back through the process to be further refined. As a result, the process leads to a mathematical object called a *fractal*. The essence of a fractal is that it is *self-similar*.

Let us suppose that as part of the process the step length was modified from stage to stage so that each resulting figure fitted into the same size square. Then, in the infinite case, if you focused on any part of the figure and magnified it to the size of the entire figure, it would be similar. The infinite version of the Snake kolam would not only be a self-similar curve; it would also be composed of four self-similar curves. In addition, although each successive curve fits into the same size square, the *lengths* of the successive curves are increasing. While the area containing them remains the same, their lengths grow exponentially.

Some of the properties of the figure place it in a special category within fractals. As we have already said, the curve is made up of a single, continuous line which never intersects itself, and which ends where it began. In mathematical terms, these properties make it a *simple closed* curve. And, since it never intersects itself and its parts never touch each other, it is a *self-avoiding* curve. In addition, still following the stipulation that the successive figures are scaled to fit into the same size square, as the number of crosses and number of arms per curve increase, the curve gets closer and closer to any point that one can specify in the square in which it is being drawn. For one of the curves in the sequence, and for every curve thereafter, the curve gets as close as is wanted to any point in the square. In the infinite case, it would fill the entire square—hence, it is called a *space-filling* curve. The combination of these characteristics leads to the finite figures being deemed FASS curves (that is, approximately space-*F*illing, self-*A*voiding, *S*imple, self-*S*imilar curves.) What is more, the angular version of the Snake kolam has been identified with a variant of the classical FASS curve known as "the Sierpiński curve" (a curve named for the mathematician Wacław Sierpiński who first discussed it in 1912), and the Snake kolam as a smoothed version of that mathematically well-known curve.

When first discussed in the mathematics literature in the late nineteenth/early twentieth century, fractals were thought very strange and even monstrous. Among the most famous early examples were the Koch "snowflake" and the Sierpiński space-filling curve. Then, due to the writings of Benoit Mandelbrot in the 1970s, the concept of a fractal was seen to provide a means of describing numerous objects occurring in nature. Since then, fractals have attracted much attention and interest. It is, at first, unexpected to find several of these same characteristics embodied in a traditional design. Upon reflection, however, we realize that it is, perhaps, the mental processes shared by humankind that result in Koch "snowflakes," Sierpiński curves, or the Snake kolam, and which are called upon to put order on what is seen in the natural world.

Through the mathematical analysis and descriptors, we are led to see more properties of the Snake kolam than we otherwise might have. Not only, for example, do we see 90° rotational symmetry and repetition, but we also see a pattern to the repetition. We can also better understand why, as we look carefully at the Snake kolam, we seem to be drawn deeper and deeper into the figure.

7 Inspired by the diversity of kolam figures, the Madras computer science group pursued new types of picture languages. Their focus was on capturing the two-dimensional nature of the kolam. A figure, such as the Anklets of Krishna, may be drawn with a single line and, hence, be defined by a string of symbols. The dot grid that precedes it and the resultant figure are none the less a two-dimensional layout. For the Anklets of Krishna, from one stage to the next, the family members are planar figures, expanding in the plane in some patterned way. Viewing the kolam as planar figures is more inclusive because it incorporates the many kolam that cannot be described by recursive string languages, or drawn with a single line. Many of these as well constitute families. While some of the families can be characterized as juxtaposed repetitions of the same basic unit, others are related by themes that are carried out through subprocesses that recur a variable number of times.

To see some families that exemplify this variety, we return to Figure 6.3. Each of the kolam in Figures 6.3a–6.3d, and the kolam in Figure 6.2g, are single representatives of families of kolam. A smaller relative

Figure 6.10 Other members of families represented in Figure 6.3. Compare parts a and b of this figure to parts a and b of Figure 6.3. (a) Mango Leaves; (b) Asanapalakai.

of the Mango Leaves of Figure 6.3a is shown in Figure 6.10a. It has a three-row arrangement of 1, 2, and 1 interior hexagons, as contrasted to Figure 6.3a which has 2, 3, and 2 interior hexagons. There are others in the family: for example, there are two five-row arrangements, one with 1, 2, 3, 2, 1 interior hexagons, and another with 3, 4, 5, 4, 3 hexagons. The kolam in Figure 6.10b is a larger version of the Asanapalakai of Figure 6.3b. While it is enlarged by an extension to the side, another, yet larger, member of this family (not shown), extends to both sides and upwards, thus containing six basic units as contrasted to the one and two units in those that are shown.

The Parijatha Creeper of Figure 6.3c belongs to still another family. As contrasted to this kolam with six leaflets (arranged as three rows of two each), another in the family has ten leaflets (arranged as five rows of two each). Two other relatives have just one leaflet in each of the first and last rows and so, in total, have four leaflets and eight leaflets, respectively. And, the Mountain Top kolam (Figure 6.2g) also comes in many sizes. The family members can be characterized by their central heights of $2n + 3$ pulli and maximum widths of $4n + 3$ pulli, for any integer n. Figure 6.2g is the Mountain Top kolam for $n = 2$; that is, it is the member of the family with a height of 7 pulli and maximum width of 11 pulli.

Thus, with this variety in mind, the Madras group developed new *array languages* in which the outcomes are *arrays* of symbols, and the rules specify how, for families of kolam, the arrays are formed. Some of the array languages are recursive, and some are not. That is, in some languages, the arrays at successive stages are developed from the array of the preceding stage, while in others, each array is developed independently, but from the same general rules. Some of their languages address rectangular arrays, which lead to dot grids, and figures, which, for all family members, retain such shapes as squares, isosceles triangles, diamonds, or hexagons. Notice, for example, the hexagonal dot layout in Figure 6.3c, the diamond-shaped layout for the Anklets of Krishna (Figure 6.7), and the essentially triangular layout of Figure 6.2g. They also considered radial dot layouts, as seen in Figure 6.3d.

In some of these languages, the individual symbols contained in an array are pictorially interpreted as two-dimensional subfigures that are assembled as specified by the layout of the symbols in the array. As an example, we return again to the Anklets of Krishna. For this, just the three symbols A, B, and C are used. The flowerlet (with dots) shown in Figure 6.6a is the pictorial representation of symbol A, a small diamond

A

(a)

B A B
A C A
B A B

(b)

B B B A B B B
B B A C A B B
B A B A B A B
A C A C A C A
B A B A B A B
B B A C A B B
B B B A B B B

(c)

Figure 6.11 Symbolic arrays for the Anklets of Krishna. (The pictorial representation of A, including the pulli, is shown in Figure 6.6a; B is interpreted as a blank; and the pictorial representation of C is a small diamond with a centered dot.) Compare parts a–c of this figure to parts a–c of Figure 6.6.

(with a centered dot) is the representation of symbol C, and a blank represents symbol B. The symbolic arrays are produced recursively. The first three of them, without the rules for moving from one to the next, are shown in Figures 6.11a–6.11c, respectively. The pictorial representations of the arrays are the drawings, including the dots, in Figure 6.6. These are the same pictures that were seen before, but they are arrived at in a different way.

Although the method just described results in figures that include the dots and, so, are closer to the actual kolam, a crucial difference still remains: in the drawing procedures of the Tamil Nadu women, the dot layouts *precede*, and in some way determine, the completion of the figures. Hence, still other, and yet more complicated, array languages were designed to reflect more closely the drawing techniques used by the Tamil women. For these, the symbols in the arrays are interpreted as dots classified into different types. The spatial layout of the dots are specified by the layouts of the symbolic arrays. Then, a small set of specific instructions, such as "join dots of one type to dots of a second type by going around dots of a third type," were given for drawing the figures.

In the development of the languages, several theoretical challenges had to be overcome. For example, the definitions of concatenation were extended and made more elaborate, and a means of overcoming the problem of shearing was introduced. Shearing occurred when, for example, the symbol A in an array was replaced by the differently configured group of symbols $\genfrac{}{}{0pt}{}{\mathrm{B}}{\mathrm{CD}}$. The latter was resolved by separating the rewriting into two phases, thus creating parallel/sequential languages.

Whatever else these languages contributed to theoretical computer science, they surely contributed by introducing the study of figures that arose in a context other than formal mathematics itself. As such, new questions were raised, and new answers had to be found. In addition, these languages focused attention on the structural variety of the kolam studied.

The final kolam family that we look at is one that is simple to draw. The members of the family are kambi kolam. They particularly attract us because, as well as appearing alone, they can be seen as building blocks in other, more elaborate kolam. Furthermore, they are similar to designs found in other religious settings, such as carvings on temple walls. The family has been described by a string language, different from the type we discussed in Sections 4–6, but one where the symbol strings are interpreted by kolam moves. The symbol string for the nth outcome, without describing the language that led to these results, is:

$$(F^{2n}R_2F^{2n}L_2)^nF^{2n}R_3(F^{2n}L_2F^{2n}R_2)^n \text{ for } n = 1, 2, 3\ldots$$

where the exponent indicates the number of times a move or set of moves is to be repeated. That is, for the first outcome ($n = 1$), the string is $(F^2R_2F^2L_2)F^2R_3(F^2L_2F^2R_2)$, and for the second ($n = 2$), it is $(F^4R_2F^4L_2)^2F^4R_3(F^4L_2F^4R_2)^2$. Figure 6.12 contains drawings of the first, second, and third stages, including the pulli that precede them.

The way this kolam family grows is interesting. Focusing on the small diamonds that make up the large interior diamond, for the successive stages, $n = 1, 2, 3$, the number of diamonds increases from 2×2 to 4×4 to 6×6. In general, the nth stage figure contains $2n \times 2n = 4n^2$ small diamonds. This mode of growth is called *polynomial* growth because it behaves like the polynomial $4n^2$. It is a *slower* mode of growth than was encountered in the Snake and Anklets of Krishna, which, in contrast, behaving like 4^n, had exponential growth.

By following with your finger the drawing steps for one or two members of this family, the next stage figure can easily be drawn, with or without reference to the symbol string. Doing this, and so experiencing the kolam kinesthetically, can help to remind us that the figures, drawn from memory on Tamil Nadu thresholds, were produced by human hand and wrist motions.

8

The kolam on the threshold are transient. Made with rice powder placed on the ground, they are walked upon and blown or swept away. None the less, the tradition evidences a number of mathematical

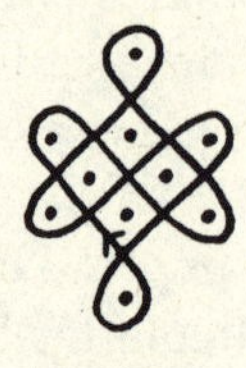

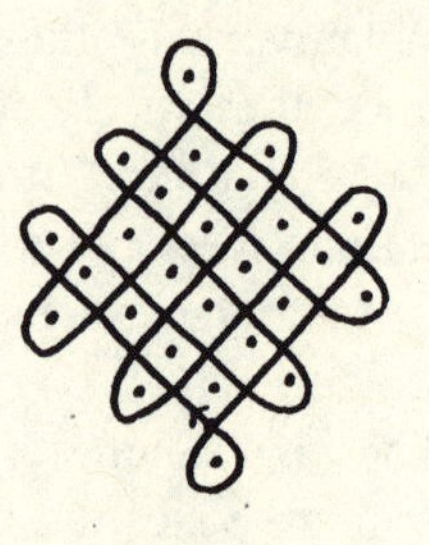

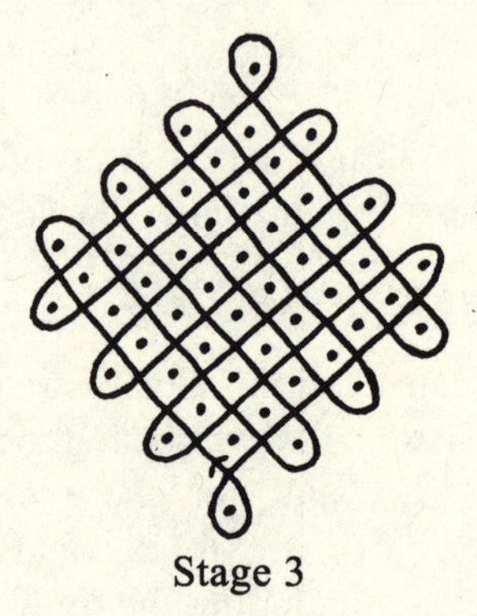

Figure 6.12 An unnamed kambi kolam. (The pulli have been added. They are not produced by the picture language.)

ideas. The kolam are diverse, and so not all of the ideas are present in all of them. Some summarizing observations can be made based on the kolam as a group.

1. For one category of kolam, the kambi kolam, the women of Tamil Nadu are interested in drawing figures continuously, ending where one began. The interest extends beyond the use of a single closed curve to the construction of figures that juxtapose several closed curves.

2. Visual symmetry is obviously of considerable importance as almost all of the kolam are, in some way, symmetric. Symmetry across a central vertical line is found in most of them. Those that do not show this symmetry generally show rotational symmetry through 90° or 180°. And in many cases, those kolam with vertical symmetry show horizontal symmetry as well.

3. Within the corpus of kolam, there are sets of figures that are visually related. Some of the figures include other figures as component

parts, or are built up by using some of the same components differently arranged. Most distinctive and most important, however, is that there are figures that are united by geometric and logical patterning. We referred to these figure groupings as *families*. The fact that the Tamil designate each of the kolam in the group by the same name is testimony that family members were considered related to each other. Even without knowing the specific techniques used by the Tamil Nadu women to remember and construct the various members of a family, it is quite probable that each member of a family was a specific expression of a general technique.

4. The first step in most of the kolam was the placement of a pulli array. These arrays had different spacings and different overall configurations depending on the kolam to be drawn. The pulli served to guide and control the drawing process, and thereby define the final figure. Thus, the learning, remembering, and drawing of kolam combined conceptualizations that were both spatial and procedural. In the kolam tradition, we see designs that are decorative, yet rich in formal structure. Furthermore, the designs are deeply embedded in Tamil Nadu culture, as they are intertwined with aesthetic and philosophical concepts, as well as with religious beliefs and practices. They are a necessary part of the knowledge and skills that women learn, and they constitute significant statements made by households on a daily basis. The underlying structures are, no doubt, an important part of the learner's ability to commit them to memory, as well as an important part of the appreciation of the figures by the community of viewers.

Notes

1. The quotation in the introductory section of the chapter is from *The King and the Clown in South Indian Myth and Poetry*, David D. Shulman, Princeton University Press, Princeton, NJ, 1985, p. 332. Pages 3–7 of the book contain valuable discussion of kolam as related to Tamil philosophy. The most extensive cultural study of kolam is "Kōlam: form, technique, and application of a changing ritual folk art of Tamil Nadu," Ralph M. Steinman, pp. 475–491 in vol. I and pp. 131–135 in vol. II of *Shastric Traditions in Indian Arts*, Anna L. Dallapicolla, ed., Steiner, Stuttgart, 1989. (My discussion of the Tamil usages of the words *kolam* and *mūli* from pp. 482–483 in this article.) Some earlier useful references are *South Indian Customs*, P.V. Jagadīsa Ayyar, Diocesan Press, Madras, 1925 (reprinted by Asian Educational Services, New Delhi, 1982), pp. 69–73, 82–88 and "Preliminary note on geometrical diagrams (kolam) from the Madras Presidency," H. Gnana Durai, *MAN*, vol. 29, May 1929, pp. 77–78.

Specifically focusing on kolam, R. Narasimhan's "The oral-literate dimension in Indian culture," pp. 67–79 in *Indological Essays: Commemorative Volume II for Gift Siromoney,* Michael Lockwood, ed., Dept. of Statistics, Madras Christian College, Madras, 1992, presents a discussion of the importance of recognizing the sophistication of oral traditions, as well as textual traditions.

Ramanujan's use of proverbs and allegories is noted in *Toils and Triumphs of Srinivasa Ramanujan the Man and the Mathematician*, Wazir Hasan Abdi, National Publishing House, Chaura Rasta, Jaipur, 1992. Pages 2–21 integrate details about Ramanujan with details about the surrounding culture. For those unfamiliar with Ramanujan, *The Man Who Knew Infinity: A Life of the Genius Ramanujan*, Robert Kanigel, Scribner, New York, 1991, can serve as an introduction.

The quoted phrase about pulli is from pp. 10–11 in "'We must make the government tremble': Political filmmaking in the South Indian state of Tamil Nadu," David B. Pratt, *The Velvet Light Trap*, Number 34, Fall 1994, pp. 10–39 (published by University of Texas Press, Austin, TX). This article provides insight into Tamil Nadu culture, as well as its uneasy relationship with India. In particular, it discusses the vibrant Tamil Nadu film industry, which, in 1980, produced over 140 films. The specific film cited here is *Ithu Engal Neethi*, written by Muthuvel Karunanidhi, and directed by S.A. Chandrasekharen.

2. The kolam in Figures 6.1a and 6.1c are reported in "Array grammars and kolam," Gift Siromoney, Rani Siromoney, and Kamala Krithivasan, *Computer Graphics and Image Processing*, 3 (1974) 63–82, and the kolam in Figure 6.1b is reported in "Labyrinth ritual in South India: Threshold and tattoo designs," John Layard, *Folklore*, 48 (1937) 115–182. Both Layard and Archana (*The Language of Symbols: A Project on South Indian Ritual Decorations of a Semi-Permanent Nature*, Crafts Council of India, Madras, 1981) report the kolam in Figure 6.1d. However, in Archana, the figure is inverted.

The kolam in Figures 6.2a and 6.2g are reported in the article by Siromoney et al. cited just above. Those in Figures 6.2b–6.2f are reported in Layard's article. Figure 6.2e is also reported in Archana.

The kolam in Figure 6.3c is reported in the article by Durai and the article by Steinman (both referenced in the Section 1 notes of this chapter). Slightly different versions of this kolam are found in Siromoney et al., and in the article by Narasimhan (cited in Section 1 notes). As contrasted to the six flowerlets, which compose the kolam shown in Figure 6.3c, the two kolam in Siromoney et al. have four and eight flowerlets, and the kolam in Narasimhan has ten flowerlets. This is an example of a *family* of kolam as discussed later in this chapter, in Sections 5 and 7. The kolam in Figure 6.3a is reported in Durai. Three different members of this kolam family are found in Siromoney et al., as are three different members of the Mountain Top family (Figure 6.2g). The kolam in Figure 6.3b is reported in Durai and in "South Indian kolam patterns," Gift Siromoney, *Kalakshetra Quarterly,* 1 (1978) 9–14. The latter also contains two additional members of this family. The kolam in Figure 6.3d is reported in Siromoney et al. (Several of the kolam in these figures were noted as reported by Layard. While I accept the validity of these figures, I am skeptical of his underlying thesis and his discussion and interpretations. In "South Indian kolam patterns" (p. 10), G. Siromoney also notes Layard's limited knowledge of the kolam, as does Archana in *The Language of Symbols*... (p. 83). Paulus Gerdes, "Reconstruction and extension of lost symmetries: Examples from the Tamil of South India," *Computers and Mathematics With Applications*, 17 (1989) 791–813, follows Layard, but even goes further to reconstruct what he believes to have been the original Tamil intentions.)

Twelve more elaborate kolam by M. Gandhimati, with an introduction by Ulrike Niklas, can be viewed on the website of the Institute of Indology and Tamil Studies, University of Cologne (www.uni-koeln.de/phil-fak/indologie/kolam/kolam1/kolamsmg.html). Also on the IITS website is a statement about kolam and why the word was selected as the title for their journal of Tamil cultural studies.

3. The commemorative volume that includes Gift Siromoney's bibliography is *A Perspective in Theoretical Computer Science: Commemorative Volume for Gift Siromoney,* R. Narasimhan, ed., Series in Computer Science, vol. 16, World Scientific, London, 1989. (A second commemorative volume, published under different auspices, has already been mentioned in the Section 1 notes above.)

I am greatly indebted to Rani Siromoney for her January 1997 letter and for sending me some articles, several references, and a 25-page original manuscript by Gift Siromoney entitled "Studies on the traditional art of kolam," Working Paper #1, May 1985. Kamala Krithivasan also was kind enough to send me a 4-page paper of her own ("Picture languages and kolam patterns"). In response to one of my questions about practioners of the kolam tradition, Krithivasan, in her February 1997 letter to me, noted that she, herself, draws these patterns in the courtyard in front of her house.

Two articles by Rani Siromoney present comprehensive technical overviews of the work of the Madras group and its relationship to other picture language studies. They are "Array languages and Lindenmayer systems—a survey," pp. 413–426 in *The Book of L*, Grezegorz Rosenberg and Arto Salomaa, eds., Springer-Verlag, Heidelberg, 1986, and "Advances in array languages," pp. 549–563 in *Graph Grammars and Their Application to Computer Science*, Harmüt Ehrig, Manfred Nagl, Grezegorz Rosenberg, and Azriel Rosenfeld, eds., Lecture Notes in Computer Science, #291, Springer-Verlag, Heidelberg, 1987. The former contains an extensive bibliography, including about 100 references to the work of the members of the Madras group.

Two additional articles that specifically involve kolam drawings and were particularly useful are Gift Siromoney's *Perception of Structure and Complexity in South Indian Kolam Patterns*, Scientific Report #62, Department of Statistics, Madras Christian College, 1986, and "Kambi kolam and cycle grammars," G. Siromoney, R. Siromoney, and T. Robinson, pp. 267–300 in the World Scientific commemorative volume noted above. (This latter article is a revised and lengthened version of "Rosenfeld's cycle grammars and *kolam*" by G. Siromoney and R. Siromoney, in *Graph Grammars and Their Application to Computer Science*, fully referenced above.)

4. For a very brief statement of the aim and approach of his early work on formal grammars, see Noam Chomsky's *Syntactic Structures*, Mouton, The Hague, 1957, pp. 13–17, 109–110.

An excellent introduction to L-systems and their interpretation as turtle graphics is *The Algorithmic Beauty of Plants*, Przemyslaw Prusinkiewicz and Aristid Lindenmayer, Springer-Verlag, New York, 1990, pp. v–vii, 1–18. Overall, this is a beautiful and fascinating book that is highly recommended. Another very good introduction is chapter 8 (pp. 1–65) in *Fractals for the Classroom*, Part two, Heinz-Otto Peitgen, Harmüt Jürgens, and Dietmar Saupe, Springer-Verlag, New York, 1992 (published in cooperation with the NCTM-National Council of Teachers of Mathematics). The concept of turtle graphics itself is presented in Seymour Papert's *Mindstorms: Children, Computers, and Powerful Ideas*, Basic Books, New York, 1980.

5. Kambi kolam are specifically discussed in the article by Siromoney, Siromoney, and Robinson, and in the working paper by Gift Siromoney, both cited in the notes for Section 3. The latter, as well as the article by Narasimhan (see notes for Section 1), discuss the procedures that were found to be used by Tamil Nadu women.

The definition of the kolam moves and their use in producing the Anklets of Krishna are in the article by Siromoney, Siromoney, and Robinson. Discussion of the production of an angular version, followed by the use of splines for smoothing, is in "Applications of L-systems to algorithmic generation of South Indian folk art patterns and Karnatic music," Przemyslaw Prusinkiewicz, Kamala Krithivasan and M.G. Vijayanarayana, pp. 229–247 in the commemorative volume edited by Narasimhan (see the notes for Section 3).

The theorem from graph theory (Euler's theorem) is discussed in chapter 2 in my book *Ethnomathematics: A Multicultural View of Mathematical Ideas*, Chapman & Hall/CRC, paper ed. 1994 in relation to the sand-drawing tradition of the Malekula of Vanuatu. That discussion also enlarges on the distinction between the tracing procedures used by the Malekula men and what *we* see in completed figures.

6. FASS curves are briefly discussed in the book by Prusinkiewicz and Lindenmayer (see notes for Section 4), and in much greater detail in an article by them and F. David Fraccia: "Synthesis of space-filling curves on the square grid," pp. 341–366 in *Fractals in the Fundamental and Applied Sciences* (proceedings of the First IFIP Conference, June 1990), Heinz-Otto Peitgen, José Marques Henriques, and Luis Filipe Penedo, eds., North-Holland Press, Amsterdam, 1991. The latter includes a discussion of the generation of a variant of the Sierpiński space-filling curve, and ends by showing the Snake kolam as an example of a smooth space-filling curve. It notes, however, that although, intuitively, the curve is self-avoiding, the extension of the definition to include smooth curves is still an open question. (Although the article refers to the curve as the Sierpiński curve, it is a variant of it. The difference is that all line segments within the curve are of equal length, as contrasted to Sierpiński's original curve in which some of the lengths differ.) Figure 6.8 is reprinted from p. 364 of this article with permission from Elsevier Science. The article also includes the angular and smoothed versions of the Anklets of Krishna. (The picture language to produce the Snake is also discussed in the last article cited in the notes to Section 5.)

The Koch snowflake and Sierpiński space-filling curve, although both commonly used as examples of fractals, were first discussed in 1904 and 1912, respectively, prior to the coining of the word "fractal" by Mandelbrot in the 1970s. It should be noted that discussions of fractals often also use other examples associated with the name Sierpiński, such as the Sierpiński arrowhead, carpet, and gasket. For a general introduction to fractals, see *Fractals: Endlessly Repeated Geometrical Figures*,

Hans Lauwerier (English translation by Sophia Gill-Hoffstädt), Princeton University Press, Princeton, NJ, 1991, and for more about space-filling curves and their history, see *Space-Filling Curves*, Hans Sagan, Springer-Verlag, New York, 1994.

7. The array languages are discussed in the two articles by R. Siromoney cited in the notes to Section 3, and in Patrick S. Wang's "Sequential/parallel matrix array languages," *Journal of Cybernetics*, 5 (1975) 19–36. Extensions to specially shaped arrays are discussed in, for example, "Hexagonal arrays and rectangular blocks" and "Radial grammars and radial L-systems," both by the Siromoneys. Both articles are in *Computer Graphics and Image Processing*, the former on pp. 353–381 of vol. 5(1976), and the latter on pp. 361–374 in vol. 4 (1975). In terms of providing specific details of kolam-producing array languages, the most comprehensive reference is the article by the Siromoneys and K. Krithivasan cited in the Section 2 notes.

The symbol string for the final kambi kolam, and the language producing it, are discussed in the article by the Siromoneys and T. Robinson cited in the Section 3 notes. The similarity of the $n = 1$ kolam to other traditional designs is discussed in G. Siromoney's "South Indian kolam patterns" (see Section 2 notes). More elaborate kolam using members of this family as building blocks are seen, for example, on pp. 137, 140, and 149 in Layard's article (see Section 2 notes).

CHAPTER 7

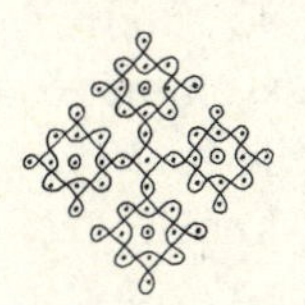

Epilogue

1 The mathematical ideas presented in the foregoing chapters are spread over time, space, and cultures. They have also involved a variety of materials. Rather than marks on paper, the ideas have been expressed through seed arrays, lines in the dust on a tray, palm ribs tied together with coconut fibers, incised wooden blocks, inscribed stone monuments, or rice-powder configurations on the ground. These are not the usual stuff of mathematics, but, clearly, the medium itself is not the central focus. What concerns us is how it is used.

In the discussion of two-valued logic in Chapter 1, we moved from the shapes 1 and 0 printed on paper, to electricity flowing and not flowing in a circuit, to one seed and two seeds in sikidy. What matters is that there are two distinct signs and that there are rules determining which of the two will result when they are combined in specific ways. From a mathematical point of view, it is just as acceptable to use one seed and two seeds for the signs as it is to use two distinct shapes printed on paper or the presence or absence of electricity. As physical objects, the stick charts of the Marshall Island navigators (Chapter 4), when contrasted with diagrams printed on paper, may be heavier to carry or seem cumbersome to us in other ways. And, of course, we are less familiar with the visual conventions of such charts. But it is their role as explanatory models, and the framework underlying the Marshallese wave piloting system that the charts embody, that are of primary significance. Furthermore, despite the fact that these and the other Marshallese stick charts that are maps were, at first glance, classed together solely because of their material likeness, the models and maps clearly differ in concept and in content.

Although different materials may have different limitations, and we may be more conscious of their limitations when they are unfamiliar, the materials can also harbor different potentials. For example, the Inca *quipus*, which we have discussed at length elsewhere, are assemblages of colored, knotted cords encoded using a sophisticated logical-numerical system. Cord colors, the relative placement of cords, knot types, and the relative placement of the knots, are all a part of the particular symbolic statement on each quipu.

Very often, because of their historical association in Western culture, writing and literacy are considered essential for the expression of mathematical ideas. And again because of its Western occurrence, discussions of writing have, until quite recently, focused solely on writing as the symbolic representation of speech sounds. Mathematical symbols, and the way they are interrelated and spatially arranged, stand *outside* of this limited view of writing. They have, as a result, received insufficient scrutiny as a mode of communication. Similarly, diagrams and illustrations that are found interspersed throughout mathematical writings are often only viewed as subordinate to words and not as forms that have their own conventions and that have developed in their own right.

When considering the ideas of traditional peoples, we become all the more aware that we need to look beyond writing in the sense of recorded speech sounds. Even where there is writing and literacy, the interplay of the oral and written varies from culture to culture, as do the materials available and the use that is made of them.

Let us look again, for example, at the kolam tradition of Tamil Nadu (Chapter 6). Because they involve transient figures drawn on the ground, they remind us of the Malekula and Tshokwe traditions, both of which involve tracings in the sand.

Among the Malekula of Vanuatu, in the South Pacific, there is a sand-drawing tradition including a stated intent, which is carried out, of tracing each figure continuously, without backtracking, and ending where one began. The figures, and the procedures by which they are drawn, were an important part of what men taught their sons. Related to beliefs about death and after-death, knowing the figures and tracing them properly were of special importance. We are fortunate to know the details of the Malekula drawing procedures for about 100 figures, and so can see that, individually, the procedures are systematic, but, more important, we can identify larger, more general systems that underlie and unite groups of figures.

The *sona* tradition of the Tshokwe, in the Angola/Congo (Zaire) region of Africa, also involves drawings in the sand. It is part of a story-telling tradition; the figures are drawn while a story is being told. The story-tellers are men; the traditional stories and figures convey the values and mores of the culture. As contrasted with the Malekula and Tamil, the art is restricted to a special few. Among those Tshokwe figures for which the drawing procedures are known, several of them are made with a single continuous line. Visually, some of the figures are remarkably similar to a few of the least ornate single-line figures of Tamil Nadu. What is more, dot grids, which precede the drawings, are a significant part of the Tshokwe drawing process. These similarities are not a sufficient reason to infer that the Tshokwe and Tamil traditions are related, but it would be most interesting if some historical linkage were to be found.

In other writings, I have elaborated on the Malekala and Tshokwe sand-drawing traditions and the mathematical ideas evidenced by them. (References to these writings are in the section notes.) The brief mention here cannot fully convey the traditions. Clearly, however, as compared to the kolam tradition, the cultures are different, the relationship of the traditions to the cultures and their meanings in the cultures are different, the underlying structures and drawing procedures for the figures are different, and, with the exception of varying degrees of concern for continuous closed curves, most of the concomitant mathematical ideas are different. None the less, what is shared is an interest in creating symmetric planar figures that are elaborated well beyond any practical necessity. And in all of these traditions, the figures and the procedures used to draw them are considered important knowledge to be carefully learned and carefully passed on..

2 In our culture, there is a marked difference between the mathematical ideas of those who specialize in thinking about them and those who do not. This distinction, and the gradations between, also exist in other cultures. Additionally, it is interesting to observe that the ideas are differently passed on and learned depending on where they are situated in a culture.

Among the Borana, for example, although everyone lives under the Gada system (Chapter 5), it is their historians who can articulate its cyclic structure and some of its implications. Similarly, specialists, such as the Maya scribes (Chapter 3) or the Rato Nalo of the Kodi (Chapter 2), were responsible for the calendars of their groups, although the calendars affected everyone. Where the ideas are pervasive, such as those that underlie kin relations, social relations, and notions of time and of space, they are part of the web of language and culture transmitted to children as they grow and mature. Each of us, somehow, learns to perceive the world and interpret experience using the same framework as do the other members of the group in which we are born and raised. Other ideas, such as those involved in the figures and procedures of the kolam tradition, or the figures and procedures of the Malekula sand-drawing tradition, are specifically taught by mothers to daughters or by fathers to sons. While some people may be more skilled and knowledgeable than others, the tradition and ideas within it are broadly shared, with no particular group identified as experts or specialists. Specialties, including, for example, weaving, carving, or pottery, involve yet other ideas, such as the visualization and creation of sizes and shapes, and the conception and execution of patterned decorations. The learning of these often involve apprenticeships, but can be as varied as the cultures and the specialties themselves.

What particularly attracts our attention, however, is the *formal* training received by some specialists in some cultures. We see this mode of learning as distinctive and as having important ramifications for the mathematical ideas being learned. What we mean by formal learning is exemplified by the training of the Marshall Island navigators (Chapter 4) and the Yoruba diviners who specialize in Ifa (Chapter 1). The training is organized, separated from daily routines, and carried out by members of a professional group. (We call them professionals, rather than experts, in part because of this training, and also because they continue to share their knowledge and interact with each other.) This mode is in decided contrast to learning that is embedded in ongoing activity and which depends primarily on observation and imitation. For both the Yorba diviners and the Marshall Island navigators, we know that there are master teachers, specially selected students, and that several students may be taught simultaneously. What is taught includes general principles, as well as specific procedures. It also includes the construction and interpretation of material arrays, and there is an abundance of detail that must be committed to memory. For the divining students, there are periodic oral examinations and a final examination. Only a fraction of those who begin the lengthy training program successfully complete it. (We do not know, but would suspect, that not all succeed who are in training to be navigators.)

Perhaps the most significant aspect of this formal training is its separation from the experiential. To convey, while on land, the meeting of several ocean waves, for example, would require some characterization of selected features of the waves. That is, since the waves themselves are not present, they have to be replaced by words and/or visual renderings describing a few features of the waves that have been *extracted* as significant. This process of extraction and of dealing with the hypothetical rather than the real is a crucial part of what distinguishes the creation and use of abstract systems from other practices. While, for the navigators and diviners, any mathematical ideas involved are phrased in terms of navigation or divination, the ideas are not surrounded by the ambiance or totality of circumstances of an actual voyage or divination session. Thus, although the ideas may be in the overall context of navigation or divination, they are decontextualized from the fullness of reality by the formal learning situation, which selectively creates its own version of that reality.

A third example of this type of formal learning that we have encountered in our studies of the mathematical ideas of traditional peoples is the case of the sixteenth-century Inca quipumakers. Here, too, those who were trained were specially selected and sent from their homes to be taught in Cuzco, the Inca capital. As result, the logical-numerical system that was learned was a shared, general system that could be used for communication and which has general principles that could be applied in different settings and in different specific instances. Although three cases of formal learning situations in traditional cultures are insufficient to generalize about their occurrence or rarity or types of subject matter, we are led to suspect that this formal mode is particularly conducive to the creation and transmission of substantial *systems* of ideas.

3 Striking in the collection of examples presented here is the prominence of cyclic structures. For the Basque, cooperation takes place through time rather than at a fixed moment, and, to accomplish this, activities are conceived of as cyclically ordered. For the Borana, events at different times are linked by their association with named classes and grades that are made to cycle through time. Continuous closed curves that are used to draw some kolam are related to the never-ending cycles of life and never-ending cycles of the seasons. Calendric structures that reflect astronomical cycles are, perhaps, most familiar to us, but it is the imposition of arbitrary cycles on time that make the calendars we have

discussed such distinctive creations of the cultures that conceived them. For the Jews, specially observing the seventh day in a 7-day cycle is of paramount importance; for the Maya and Balinese, the significance and quality of any day is determined by the particular set of gods that influence that day as a result of the interaction of multiple cycles. In all of these systems, with the exception of the kolam, there are finite length cycles of discrete elements. Crucial to the behavior of these systems are the cycle lengths, which vary considerably, including, for example, five and seven; ten and six; twenty, eighteen, and thirteen; and one, two, three, four, …, nine, and ten.

Given the ubiquity of cycles in structures created and imposed by human beings, it is not surprising that cyclic structures have been well studied by our mathematicians and that arithmetic and algebraic rules for dealing with these structures are available to us. Others may have used different approaches, but they did, none the less, deal with cycles and with cycles combined with other cycles.

4 Sources of information about the mathematical ideas of traditional peoples have been scant and not of the sort mathematicians or historians of mathematics are used to. Until recently, over 90% of traditional cultures had no writing as we generally use the term. They, therefore, left no documents about their ideas expressed in their own words. To learn about their past ideas and traditions, we often must depend on information that can be extracted from artifacts or from reports of observations left by others. Too often, these others did not fully understand what they saw and were not especially concerned with mathematical ideas. Not atypical are the early report of sikidy in Madagascar by a traveler in the 1600s or the first report of the Marshall Islands stick charts by a missionary in the mid-1800s. Until the late 1800s/early 1900s, there are primarily reports by missionaries, travelers, sailors, and sea captains. Then, European ethnologists traveled widely, amassing details about the cultures they had heard existed. The lens through which they viewed these cultures was, as we might expect, their own European culture of the time, with all of its assumptions and prejudices. Other than, perhaps, listing some number words, mathematical ideas were unanticipated, unsought, and generally overlooked. However, what there is can be very useful, because contained therein are lots of details, including drawings and photographs, which can be reviewed from a mathematical perspective. Many artifacts that are now in museums were collected as part of these turn-of-the-century expeditions.

But, as we noted in the introduction, during the past 80 years, much has been learned about culture, about language, and about cognitive processes. Not only can these new understandings be brought to bear on earlier data, but later studies by ethnologists, linguists, and cognitive scientists, such as many of the studies done after 1950 or 1960, are already framed by these twentieth-century theories and sensibilities. Most of these later studies include more focused examinations or re-examinations of ideas and traditions that continue. They also include an appreciation of the intellectual capacities and knowledge of traditional peoples, as evidenced by the emergence of such fields as ethnobotony, ethnoastronomy, or ethnoichthyology. In general, these fields focus on a culture's conceptualizations, perceptions, beliefs, and activities related to plants, celestial bodies, or fish, respectively. But, in part, because of the entrenched view that mathematical ideas are culture-free or culture-neutral, it was not until even more recently that mathematical ideas were also studied in this way. Among these recent sources are some particularly fruitful studies due to scholars in such fields as archeology, ethnology, linguistics, and culture history, who have become cognizant of the need to pursue mathematical ideas, and do so using their expertise and their methods.

One example of a recent investigation by an anthropologist is a study done among contemporary Quechua-speaking people of the Andes. The study focuses on the conceptualization, use and meaning of numbers in their culture. What emerges is, in the words of the investigator, "a unique ontology of numbers and philosophy of arithmetic..." that is new to us and stands in decided contrast to the view of numbers put forth by some Western philosophers of mathematics. For mathematicians, however, this is not the first time that counting and numbers in other cultures have been discussed or contrasted with the ideas of Western philosophers. But this extended study in itself substantially challenges the idea that the conception of numbers is culture-free or culture-neutral. The study also underscores that the use of numbers is evidenced in different contexts in different cultures. A significant part of the study discusses numbers and counting in the context of weaving, an activity which is of central importance in Andean culture. In this instance, as in so many others, the author and his collaborator's knowledge of the language play a crucial role, since it is the language that carries the ideas. Too often when relying on reports of outsiders not specifically interested in mathematical ideas, or not sufficiently acquainted with the language, translation has already unintentionally served to filter or recast the ideas so that they are overly simplified or sound more or less like our own.

It is important, however, to recognize that the beliefs discussed in the study are those of the Quechua-speaking population at large, rather than the ideas of some few specialized professionals in their culture. No doubt, a careful study in our own culture would similarly show that the concepts of the population at large diverge considerably from those of our philosophers. In fact, numerous recent writings by professional working mathematicians question whether the philosophers even speak for them. Thus, as we broaden our vision to include other cultures, we must also keep a mind that in any culture, whether our own or that of others, there are some ideas that pervade the culture, some ideas that are particular to specialized groups within the culture, and some ideas that are special to specific individuals. Acknowledgment and understanding of their differences, similarities, and how they fit together could create a more nuanced history of the mathematical ideas in any culture, including our own.

An idea that has attracted much interest and led to a variety of different studies, large and small, is the creation and use of repeated designs arranged along strips or in two-dimensional arrays. This formal mode of spatial decoration has been found to be unusually common in cultures spread through space and time. In addition, independent of concern for the ideas of other cultures, there has been mathematical interest in the use of symmetry groups for the analysis of such strip and surface patterns. These two streams of interest merged as long ago as the 1940s when, for a Ph.D. thesis, formal mathematical analysis was applied to Moorish decorations in the Alhambra, and, just a short while later, using Pueblo pottery as an example, an archeologist advocated this mode of analysis for pottery classification. Now, particularly in educational settings, this mode of analysis has become even more widespread, as it enables the inclusion of materials from other cultures. Through the use of collections of realistic patterns, there can be seen different expressions of the same mathematical abstraction as they arise in meaningful human settings. The collections of patterns clearly show that despite differences in style, context, meaning, and materials, the same formal spatial orderings occur in many different cultures. Although there is no implication (or there should not be) that our mode of analysis or the resulting categorization into symmetry classes expresses the ideas of the patterns' creators, several studies of small collections focus on whether or not various cultural groups used all the symmetry classes that were possible. Other studies of larger collections focus on which symmetry classes were most prevalent in the patterns used by a people.

An important complement to this analytic approach is a recent field study of the rafia-cloth design categories utilized by the Bakuba who live in the Congo (Zaire) region of Africa. Mathematical analysis of the Bakuba material itself first appeared about 30 years ago in the *Journal of Geometry*, and then again soon thereafter in the seminal *Africa Counts*. The recent Bakuba study is by an archeologist who had in the past collaborated with a mathematician on symmetry studies. The goal of the field study was to find out what kinds of properties or features contemporary Bakuba actually use to define a design category. Here, again, language is crucial as naming and the visual identification of what is named are intimately linked with each other. As we look at the numerous designs and read about how the Bakuba see them, we are again made aware that the shapes and juxtapositions we see as basic—and as distinguishing some designs from others—are neither universal nor objective. They are part of our perceptual framework and not necessarily what is seen by others. Similarly, some of the distinguishing features that the Bakuba clearly see are not necessarily at all clear to us. As we would expect, the features that we use to define the symmetry classes of the two-dimensional arrays with patterned repetition are not the features that concern the Bakuba. Nevertheless, the Bakuba samples analyzed show consistently greater use of ten of the seventeen possible symmetry classes.

Another recent study, quite different in kind, is about the previous septuagesimal system of units of the Basque. The author, whose special area is Basque studies and ethnoastronomy, argues that the Basque metrological tradition depended on a cognitive framework that differed considerably from that commonly associated with Western mathematics. The framework was an integral part of the Basque cosmovision as expressed in folk tales and ritual performances, as well as in celestially encoded star figures. The origin of the septuagesimal system is still unknown as some aspects of the Basque system are also found among their former Celtic-speaking neighbors and even among the ancient Greeks. For the Basque, the ramifications of the septuagesimal system extend from the placement, some 2000 years ago, of stone octagons to identify pasturage, to a calculating device used in navigation and map plotting in the Middle Ages. For us, the study is particularly significant as it makes us realize that even the histories of European science and European mathematics are more culturally variegated than we generally think of them, extending beyond just differences in practices to differences in the conceptualization of space and time.

Recognition that there exist, in traditional cultures, mathematical ideas more interesting and substantial than was previously believed has been an important first step in these and other recent studies. As this recognition grows and spreads, we anticipate additional studies. And, just as we noted that many people now live under more than one calendar, a growing number of people are living with more than one cultural tradition, namely, the culture in which they were born and raised, and the culture whose idea dominated their formal education. By retaining and valuing both traditions, they will, we believe, make special contributions as they raise new questions, seek new answers, and develop new perspectives. We already see in the discussions of the Gada system (Chapter 5) and of the kolam tradition (Chapter 6) significant contributions by scholars who combined their own cultural understandings with what was learned in academic settings dominated by Western ideologies.

5 For all of us, on whatever level of learning, knowledge of the ideas of others can enlarge our view of what is mathematical and, in particular, add a more humanistic and global perspective to the history of mathematics. This enlarged view, in which mathematical ideas are seen to play a vital role in diverse human endeavors, provides us with a richer and fuller picture of mathematics and its past.

Twenty-first century mainstream mathematics is reaching people of more and more diverse cultures as the teaching of it continues to spread across national and continental boundaries, as people move from one country or region to another, and as several cultures are represented in the backgrounds of more individuals. An enlarged view of the past can help in furthering the realization that people of different cultural traditions will enrich mathematics itself by bringing to it different perspectives and different ways of perceiving and categorizing the world.

Notes

1. Rievel Netz's *The Shaping of Deduction in Greek Mathematics: A Study in Cognitive History,* Cambridge University Press, Cambridge, 1999, is highly recommended. The book is exceptional in that it discusses the role and use of diagrams in Greek writings, focusing on the period from the middle of the fifth century BCE

to the middle of the fourth century CE. Diagrams and text are seen as interdependent, with diagrams one of the two main tools that shaped the Greek method of deduction. The significance of cultural context is emphasized throughout. The style of Greek mathematics is viewed as resulting from their combination of writing with their tradition of orality.

Also highly recommended is R. Narasimhan's "Literacy: Its characterization and implications," pp. 177–197 in *Literacy and Orality*, David R. Olson and Nancy Torrance, eds., Cambridge University Press, Cambridge, 1991. In discussing literacy, he includes the perspective of the Indian tradition as contrasted with solely using the Western tradition. He argues that literacy must be viewed more broadly than just alphabetic literacy and notes the insufficiency of writing in the engineering/engineered worlds. The article is particularly important because it extends consideration to diverse forms of symbolic notation. It also touches on abstraction, the construction of formal models, and two-dimensional spatial representations. (Also see Narasimhan's article "The oral-literate dimension in Indian culture" cited in the Section 1 notes of Chapter 6.)

Inca quipus are discussed in detail in Marcia Ascher and Robert Ascher, *Mathematics of the Incas: Code of the Quipu*, Dover Publications, New York, 1997. Also, my article "Reading quipus: Labels, structure, and format," in *Narrative Threads: Explorations of Narrativity in Andean Khipus*, Jeffrey Quilter and Gary Urton, eds., University of Texas Press, in press, focuses on their logical-numerical system as a symbolic system that does not encode speech sounds. Quipus are placed within a broadened conception of writing in the article "Inca writing" by Robert Ascher, in the same volume.

The Malekula and Tshokwe sand-drawing traditions are discussed in Chapter 2, "Tracing graphs in the sand," in my book, *Ethnomathematics*, which is cited in full in the notes to Section 5 of Chapter 6. More detailed discussions of them, also by me, are in "Graphs in cultures: A study in ethnomathematics," *Historia Mathematica*, 15 (1988) 201–207 (Malekula), and "Graphs in cultures (II): A study in ethnomathematics," *Archive for the History of Exact Sciences*, 39 (1988) 75–95 (Tshokwe). Paulus Gerdes has written extensively about the Tshokwe tradition, primarily suggesting classroom activities inspired by the material. See, for example, Chapter 4, "The 'sona' sand drawing tradition and possibilities for its educational use," in his *Geometry from Africa: Mathematical and Educational Explorations*, Mathematical Association of America, Washington, DC, 1999. Of his writings, the most informative about the tradition itself is *Sona Geometry: Reflections on the Tradition of Sand Drawings in Africa South of the Equator*, vol. 1, Instituto Superior Pedagógico, Mozambique, 1994 (translated into English by Arthur B. Powell of Rutgers University, Newark, NJ).

2. The training of the Yoruba diviners is discussed in detail, as a case in point, in Festus Niyi Akinnaso's "Schooling, language, and knowledge in literate and nonliterate societies," pp. 339–385 in *Cultures of Scholarship*, S.C. Humphreys, ed., University of Michigan Press, Ann Arbor, MI, 1997. The focus of the article, however, is formalized learning. In it, Akinnaso challenges the view that formalized learning is necessarily associated with the transmission of literate knowledge, as contrasted with socially embedded learning in nonliterate societies. This insightful article contains numerous references and is especially recommended to those concerned with educational practices and with bridging the gap between practical knowledge and specialized knowledge.

Chapter 4 of *Mathematics of the Incas*, fully cited in the Section 1 notes above, is devoted to the quipumaker. It also includes references to the education of Sumerian and Egyptian scribes. For another detailed discussion of the training and role of the Old Babylonian scribes, see "Mathematics and early state formation, or the Janus face of early Mesopotamian mathematics: Bureaucratic tool and expression of scribal professional autonomy," pp. 45–87 in *In Measure, Number, and Weight: Studies in Mathematics and Culture*, Jens Høyrup, State University of New York Press, Albany, NY, 1994, in particular pp. 64–66 and 82–84.

4. The study done among contemporary Quechua-speaking people in the Andes is discussed in chapters 1–5 of *The Social Life of Numbers: A Quechua Ontology of Numbers and Philosophy of Arithmetic*, Gary Urton with the collaboration of Primitivo Nina Llanos, University of Texas Press, Austin, TX, 1997. (As the author himself notes, chapters 6 and 7, which attempt to connect these ideas to pre-Hispanic and colonial Andean societies, are less successful.) The phrase quoted from Urton is on page 3 in his book. To place this study in a broader context, see, for example, the chapter devoted to number words and number symbols in my book *Ethnomathematics*.

Symmetric strip decorations are the subject of Chapter 6 in *Ethnomathematics*. In it, I discuss the mathematical analysis of one- and two-color strips, as well as examples from Inca pottery and carved wooden Maori rafters. Numerous references are contained in the notes to that chapter, including references to the early work by crystallographers and mathematicians, the 1944 Ph.D. thesis, and the 1948 book by the archeologist Anna O. Shepard.

Examples of recent collections by mathematics educators are *The Algebra of the Weaving Patterns, Gong Music and Kinship System of the Kankana-ey of Mountain Province*, Faculty of the Discipline of Mathematics, University of the Philippines, College Baguio, 1996; "Designs and patterns," section 8.4 (pp. 164–179) in *Fijian Perspectives in Mathematics Education*, Salanieta Leiloma Bakalevu, Ph.D. thesis, University of Waikato, Hamilton, New Zealand, 1998; "Symmetry patterns of the Wisconsin woodland Indians," Kim Nishimoto and Bernadette Berken, *International Study Group on Ethnomathematics Newsletter*, vol. 12, no. 1, November 1996, pp. 6–8; and *Sipatsi: Technology, Art and Geometry in Inhambane*, Paulus Gerdes and Gildo Bulafo, Universidade Pedagógica, Maputo, Mozambique, 1994.

Analysis of the Bakuba patterns appeared in "The geometry of African art I. Bakuba art," *The Journal of Geometry*, 1 (1971) 169–181, by Donald W. Crowe and then in a section on symmetry analysis (pp. 190–196) written by him in Claudia Zaslavsky's *Africa Counts*, Prindle, Weber, and Schmidt, Boston, MA, 1974. (A twenty-fifth anniversary third edition of *Africa Counts* is available from Lawrence Hill Books, Chicago, IL.) The recent Bakuba study is *Style, Classification and Ethnicity: Design Categories on Bakuba Raffia Cloth*, Dorothy K. Washburn, Transactions of the American Philosophical Society, vol. 80, part 3, The American Philosophical Society, Philadelphia, PA, 1990. (Dorothy K. Washburn and Donald W. Crowe collaborated on *Symmetries of Culture: Theory and Practice of Plane Pattern Analysis*. University of Washington Press, Seattle, WA, 1988.)

The Basque study is discussed in "An essay on European ethnomathematics: The coordinates of the septuagesimal cognitive framework in the Atlantic Facade," Rosyln M. Frank, 78 pp., ms., May 1995. Part of this, in revised form, has appeared as "An essay on European ethnomathematics: The Basque septuagesimal system. Part I," pp. 119–142 in *Actes de la Vème Conférence Annuelle de la SEAC*, Arnold Lebeuf and Mariusz S. Ziólkowski, eds., Départment d'Anthropologie Historique, Institut d'Archéologie de l'Université de Varsovie–Musée Maritime Central, Warsaw, 1999. The stone octagons are discussed in "The geometry of pastoral stone octagons: The Basque *sarobe*, Rosyln M. Frank and Jon D. Patrick, pp. 77–91 in *Archeoastronomy in the 1990s*, Clive L.N. Ruggles, ed., Loughborough Group D Publications, London, 1993.

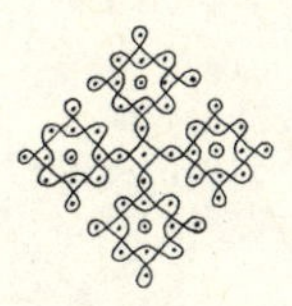

Index